AF531129

OPTOELECTRONICS GUIDEBOOK– with tested projects

OPTOELECTRONICS GUIDEBOOK– with tested projects

By R. W. Fox

TAB BOOKS
Blue Ridge Summit, Pa. 17214

FIRST EDITION

FIRST PRINTING—JANUARY 1977

Printed in the United States
of America

Hardbound Edition: International Standard Book No. 0-8306-6836-5

Paperbound Edition: International Standard Book No. 0-8306-5836-X

Library of Congress Card Number: 76-45067

Library of Congress Cataloging in Publication Data

Fox, Robert W 1934-
Optoelectronics guidebook.

Includes index.
1. Optoelectronic devices. I. Title.
TA1750.F69 621.36 76-45067
ISBN 0-8306-6836-5
ISBN 0-8306-5836-X pbk.

Foreword

There are currently few books dedicated to the area of optoelectronics, though several of the optodevices have been available for many years. The introduction of low cost LEDs has broadened this field considerably. To match the semiconductor light emitters, the industry has also expanded the semiconductor detector lines, making possible applications that were uneconomical only a few years ago.

This book is designed primarily for the hobbyist and technician. It contains the basics of how most optoelectronic devices work, without several pages of equations. The projects have all been built and tested and will work when properly built by the reader. It is not possible to foresee all the possible substitutions of parts or errors made while constructing these projects. The author and the editors, therefore, can assume no liability for the construction or use of these projects. In addition, some of the circuits contained in this book may be covered under patents.

The author wishes to thank Neal Cleare of International Rectifier and David Seefeld of the General Electric Company for their cooperation in the development of this book.

R.W. Fox

Contents

Introduction To The Projects

Electronic circuits can be fun to build. They can also be very frustrating. The best way to insure success is not to tackle projects that are too complex for you. The second factor is to be systematic. Don't jump from project to project, but finish one before starting the other. The third and probably most important, use quality components and equipment. One bad component in a circuit may take hours to locate, and a corroded soldering iron tip may cause you to destroy several times its value in components.

As an aid to the builder of these and other projects, remember that Chapter 6 contains a detailed troubleshooting guide. If all else fails, reread the instructions.

Good luck and enjoy the projects!

Chapter 1 Characteristics of Light

Optoelectronics is a portion of electronics which deals with the interaction of light with electronic circuitry. There are, as we will see within this book, electronic devices which change their characteristics when exposed to light and there are other electronic devices which generate light. The first task before us is to understand light, what it is, and how to control and operate with it.

LIGHT

Light can be considered as a part of the electromagnetic spectrum as shown in Fig. 1-1. We, for the purpose of this book, define light as any electromagnetic energy in the ranges of the ultraviolet, visible, or infrared portion of the electromagnetic spectrum. The units of wavelengths commonly used will be angstroms or microns. An angstrom is 1/10,000,000 of a centimeter, while a micron is 1/10,000 of a centimeter. The actual division among the portions of the electromagnetic spectrum is arbitrary. For example, there have been shortwave radio transmissions that are of a shorter wavelength than the infrared radiation emitted by some hot objects. We also run into problems defining the efficiency of the electromagnetic radiation sources because, as we know,

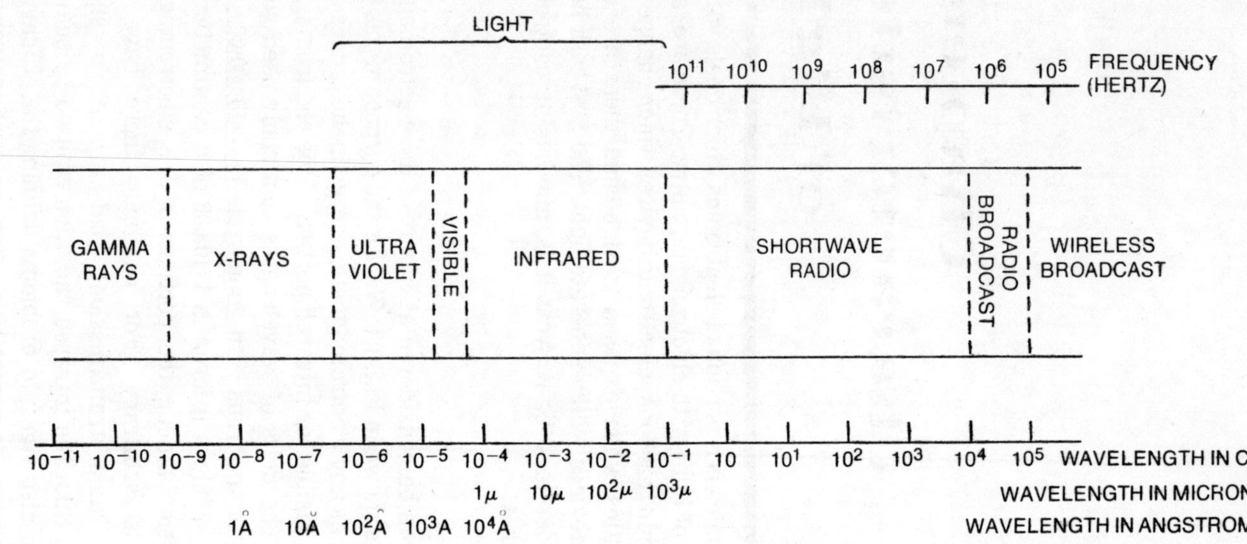

Fig. 1-1. The electromagnetic spectrum covers wavelengths from the very long to the very short. Visual light covers only a very narrow range of this spectrum.

energy cannot be destroyed, only transformed. Therefore, the efficiency of an electromagnetic radiator is defined in relation only to the radiation in the desired portion of the spectrum.

The speed of light in a vacuum or air is about 3×10^{10} cm/sec. It is at this speed that all electromagnetic energy is radiated in air or a vacuum. From this we can characterize the frequency of light as the speed of light divided by the wavelength. Experiments have shown that as the light source weakens, the energy in the radiation does not decrease linearly but in discrete steps, indicating that for a given wavelength (or frequency) there are finite bundles of energy. The energy of the bundle, which is called a photon, is proportional to the wavelength of the energy by a constant known as Planck's constant, after Max Planck, the originator of quantum physics. Planck said that light waves are not continuous streams of energy, but discrete bundles of energy (photons) and have a characteristic frequency of oscillation.

GEOMETRICAL OPTICS

The oldest branch of optics is geometrical optics. Though it is very simplistic with respect to the discussion above, it is a convenient way to study lenses, mirrored surfaces, and the like. In optoelectronics, it is usually necessary to use these devices in order to obtain optimal performance.

In geometrical optics the concept of a light ray is important. A light ray is a very narrow light beam; so narrow it is only a line in space. The ray shows us the direction of travel of light energy. Figure 1-2 shows a light ray (I) striking

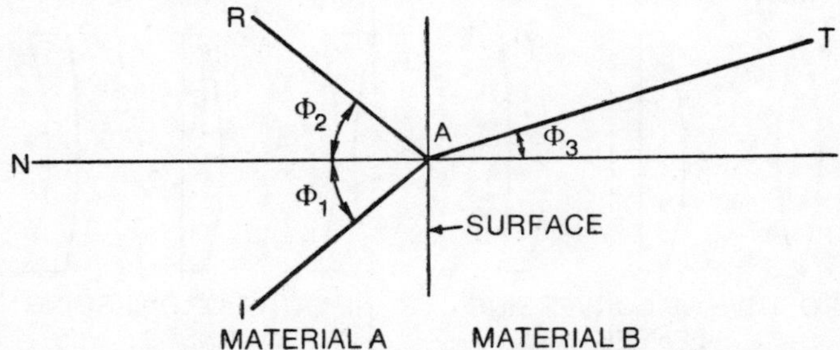

Fig. 1-2. Light rays striking a surface are either reflected (IAR) or refracted (IAT). The angle of the reflected ray (RAN) is identical to the incident ray (IAN). The refracted ray's angle depends upon the two media.

the surface of a material at A. If the material is transparent, the light will generally be divided into two rays. One is the reflected ray (R) and the other the refracted ray (T). The reflected ray bounces off the surface while the refracted ray is bent going past the surface. The angles between the incident ray and reflected rays are equal. That is, if N is a perpendicular to surface angle IAN, then surface angle IAN is equal to angle RAN.

The refracted light ray's angle to the perpendicular is also a function of the angle at which the incident ray strikes the surface. In this case the two angles are related as a function of the index of refraction of the two materials as:

$$\frac{\sin(\text{IAN})}{\sin(\text{TAN})} = \frac{\sin(\Phi_1)}{\sin(\Phi_3)} = \frac{N_A}{N_B}$$

where N_A and N_B are the index of refraction of the two materials. The index of refraction of air or a vacuum is 1.0. In all other materials the index of refraction is higher than 1.0.

A third law of geometrical optics, which is important to our understanding here, is the principle of reversibility. This is that light rays traveling in the opposite direction will follow the same path. Therefore, if in Fig. 1-2, TA was the incident ray, then AI would be the refracted ray. Likewise with RA being the incident ray, then IA would be the reflected ray. This

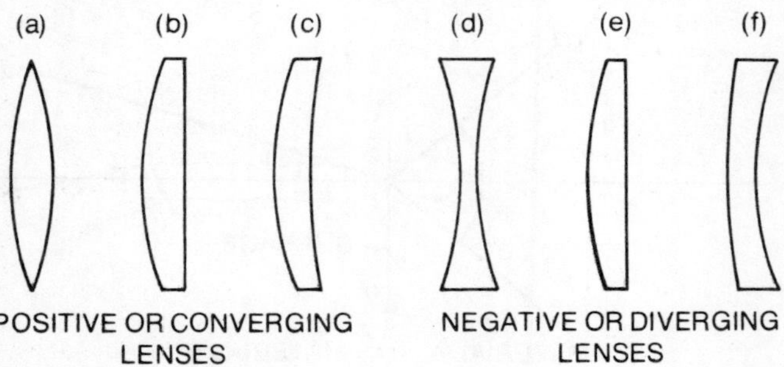

Fig. 1-3. Cross sections of some common thin lenses. (a) equiconvex (b) planoconvex (c) positive meniscus (d) equiconcave (e) planoconcave (f) negative meniscus

bending action which occurs at the surfaces allows the construction of lenses. Figure 1-3 shows several of the standard-form lenses. The converging lenses are thinner at the ends than in the center, while the diverging lenses are thicker at the edges. Convex lenses bow out at the center and the concave lenses bend inward. Plano in the lens name indicates a flat surface.

What happens when light passes into one of the lenses is shown in Fig. 1-4. Each subfigure shows light passing through one surface. Therefore, an equiconvex lens (a lens that is convex on both sides) would be represented by two of either Fig. 1-4a or 1-4c. So that from a point source of light (Fig. 1-4a) the equiconvex lens would converge light at a second point on the other side of the lens. From a parallel beam, the lens would yield a similar but smaller parallel beam on the output side. By geometrical construction the effect of any lens or combination of lenses can be shown. Mirrored surfaces can be handled in a similar manner.

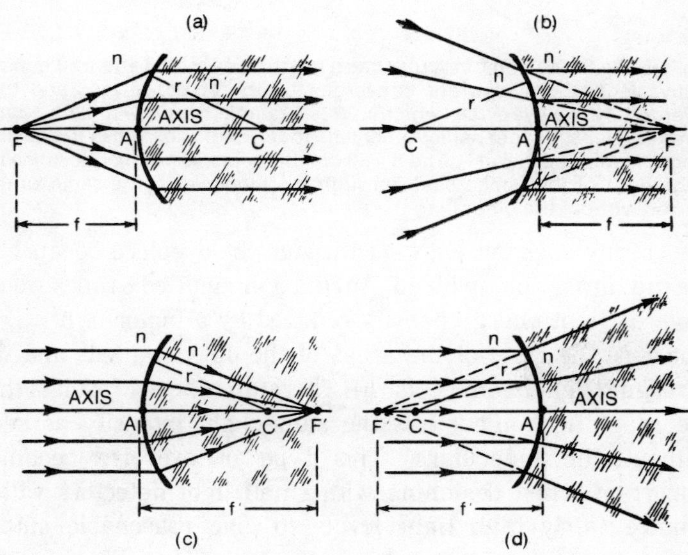

Fig. 1-4. Diagram demonstrating the effects of lenses on light rays. F and F′ are called focal points while f and f′ are the respective focal lengths. An equiconvex lens as shown in Fig. 1-3a would be two of Fig. (a) back-to-back if illuminated from a point source of light.

Figure 1-5 shows the effects of a lens system on light intensity. The lenses represent half of a standard 8 × 50 mm binocular in which the objective lens is a 50 mm diameter converging lens and the eye piece is a 10 mm diverging lens. The two are positioned such that the light coming in the objective lens exits in a 6.25 mm diameter beam. The eye pupil is about 7 mm in diameter, so the entire beam can pass into the eye. If a given light energy density is entering the objective, all

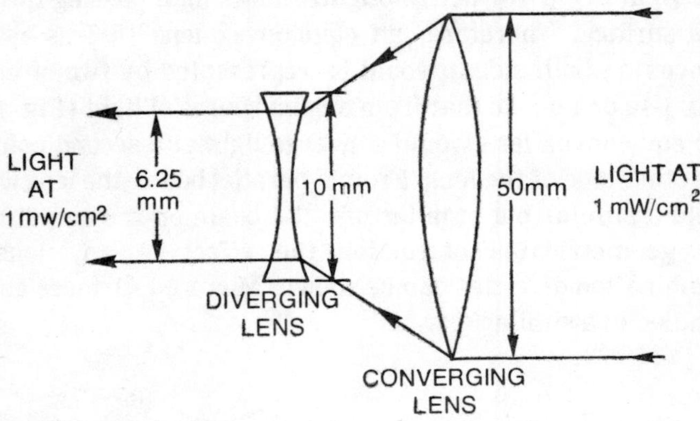

Fig. 1-5. An 8×40 mm binocular has a magnification of 8 and a 50 mm objective lens. Light rays are converged by the objective lens and then diverged by the eyepiece, which in this example happens to be a 10 mm diverging lens. Proper selection of lens focal lengths provides the desired magnification power of 8. The 8×50 mm binocular is popular because the brightness of the image seen through the binocular is the same as that viewed without the binocular.

the energy does not pass out through the eyepiece because of the magnification involved. An image magnified 8 times would have its light energy density reduced by a factor of $8^2 = 64$. But here the ratio of the areas of the objective lens and the emerging light beam is also 64. Therefore, the net result is that the magnified object has the same light intensity as seen without the binoculars. This type of exercise becomes important when designing with small light detectors which require fairly high light levels to give reasonable output signals.

PHYSICAL OPTICS

Physical optics comprises the study of phenomena which have a bearing on the nature of light. Examples are the interaction of light with matter, light with itself, and light with other forms of energy. For the purposes of this book we shall only concern ourselves with the physical phenomena of absorption and scattering.

When a light wave enters a medium, its intensity is generally reduced as it passes through the material. Some of the energy of the light is absorbed by the atoms of the material and some is redirected or scattered. Some materials affect some wavelengths differently than others. Such materials are said to show selective absorption or scattering. Consider normal white light shown through a piece of red glass. The glass has absorbed all the blue and green light and lets only the red pass through. These effects and the knowledge of the wavelengths affected are important when operating with optoelectronic systems, since most devices are sensitive only to specific wavelengths.

FILTERS

It is necessary in optoelectronic systems to know what frequencies and how much of what frequencies are present in order to guarantee correct operation of the system. Figure 1-6 shows an example of a transmitting and receiving system. The system consists of a point-light source that has a spectral output that is shown in Fig. 1-6a. The light is made into a parallel beam by the first converging lens, and is transmitted following a path containing a filter whose characteristics are shown in Fig. 1-6b. The light is collected by the second converging lens to a point at the detector. The detector's spectral characteristics are described by the curve shown in Fig. 1-6c.

Let's examine what occurs in the system. The light output is similar to several gas discharge sources used today. At "A" the light spectrum remains unchanged, but when the light passes through the filter some wavelengths are almost completely absorbed, such as the light at one micron. Other wavelengths are attenuated or partially absorbed giving a light spectrum at "B," as shown in Fig. 1-6d. The characteristics of the filter are typical of bandpass filters used in optics and are analogous to bandpass filters used in

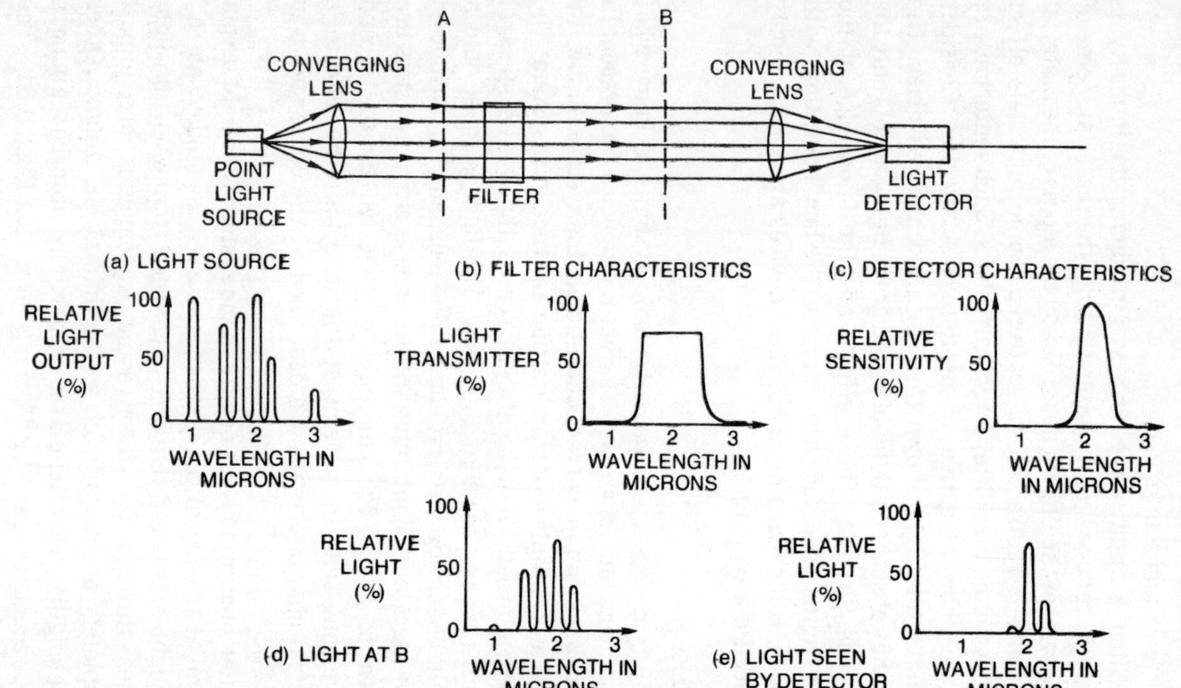

Fig. 1-6. A simple optical system showing the effects of filtering and spectral sensitivities on the effective light transmitted. The effects of each portion is explained in the text.

Table 1-1. Comparison of the Relative Sensitivities of the Eye and a Silicon Phototransistor To Different Light Sources.

Radiators \ Detectors		Human Eye	Silicon Phototransistors
Tungsten Lamp	2000°K	.003	.16
	2200°K	.007	.19
	2400°K	.013	.22
	2600°K	.021	.24
	2800°K	.030	.27
	3000°K	.044	.30
Neon Lamp		.35	.7
GaAs LED 0.9μ		0	1.0
GaP LED 0.7μ		.08	.7
SiC LED 0.6μ		.6	.16
Fluorescent Lamp		.1	.4
Xenon Flash		.13	.5
Sun		.16	.5

electronics. The light detector does not see all wavelengths, just as our eyes do not see the infrared or ultraviolet. The light that it receives is the light at "B" but it only "sees" light as shown in Fig. 1-6e. You can see that the filter blocked light that the detector was sensitive to, as well as light that the detector could not see anyway. The net result is that the usable and unusable light has been attenuated by the filter. Without the filter the detector would have seen more light.

The filter is not a discrete part of the system in many applications, but a result of the components used, such as lenses and fiber optics. Air is also a filter to certain wavelengths. Care must be taken, therefore, in building optical systems to insure that the desired wavelengths are both present at the source and not severely attenuated by the optical path.

Table 1-1 gives the relative responses of the eye and a silicon phototransistor to several different light sources as an example of the filtering effect and the effect of differences in spectral response of emitters and detectors. It is interesting to note that in every case shown the transistor sees more of the energy from the source than does the eye, and in one case, that of the GaAs LED, the silicon transistor sees all of the radiation while the eye sees none.

Chapter 2 Light-Sensitive Devices

There are many light-sensitive devices outside of electronics. The eyes are probably the best known to all of us. With our eyes we perceive the photons which enter the eye. The eye, with the brain, decodes these photons and creates within the mind a full color picture. In a brief instant many millions of bits of information are received and analyzed by the brain. This information is used by us to allow us to interact with our environment. The eye does not react to all the photons it receives. For example, a hot object emits photons in the infrared range of the spectrum. We cannot see that the object is hot unless it also emits photons in the red spectrum and, as a result, people often burn themselves on hot objects. On the other side of the visual light spectrum we cannot see ultraviolet, but our skin is very sensitive to these photons. We must depend upon a secondary measurement, the reddening of the skin, to react. Sunlight has much ultraviolet; we can all remember not reacting soon enough and receiving a bad sunburn.

The electronic light-sensitive devices are similar in that they are sensitive to only specific bands of photon energy. Figures 2-1 and 2-5 show the relative sensitivity of the different semiconductor materials which are commonly used in light-sensitive devices. The characteristics of the human eye have been included for comparison. It is obvious from Fig. 2-1 that these materials, except for cadium sulfide, are more sensitive to the infrared wavelengths than to visible light.

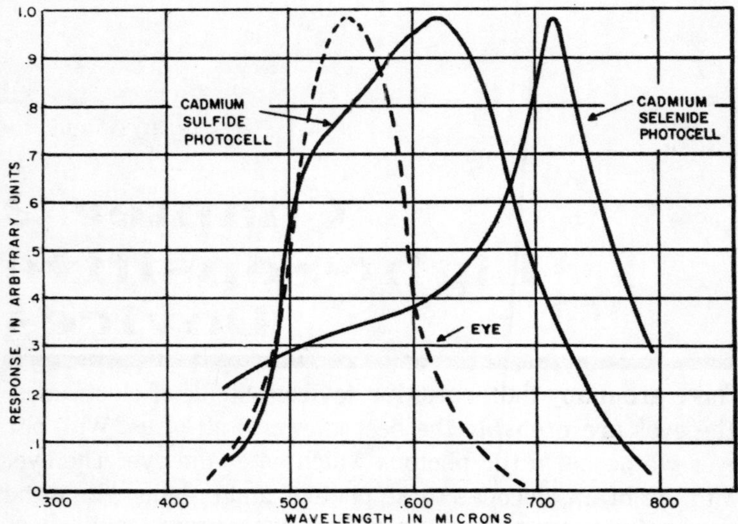

Fig. 2-1. Relative spectral response for the two basic photocell materials, cadmium sulfide and cadmium selenide. Note that cadmium sulfide is more sensitive in the same wavelengths as the eye. For this reason it is used more often in light meters. (Courtesy of General Electric, Lighting Systems.)

The semiconductor light-sensitive devices can be divided into two main categories. In the first is the photoresistive device. These are generally called photocells. The second category is the junction devices. This class is then divided into the photovoltaic devices, which convert light energy into electrical energy, and conductance-controlled devices such as the phototransistor, photo-SCR and the like, which use light to control their current conductivity characteristics.

PHOTOCELLS

Photocells, photoconductive cells, or photoresistive cells are all common names for a group of light-sensitive resistive devices. These units are analogous to a thermistor, except where the thermistor's resistance varies with heat, a photocell's resistance varies with light intensity. They are shown schematically by a resistor in a circle with a Greek lambda within the circle. The commercially available photocells are made from either cadmium sulfide (CdS), or cadmium selenide (CdSe).

When the photocell manufacturer makes a photocell, he first determines the material he wishes to use. This will depend upon the spectrum of the light source and the relationship of the application to the different material characteristics. Table 2-1 shows the relative characteristics of cadmium sulfide and cadmium selenide. For example, if higher speed of response is important, the manufacturer would

Table 2-1. Comparison of characteristics of cadmium sulfide and cadmium selenide photocells. Each has properties which are important for different applications. (Courtesy of General Electric.)

Characteristic	Cadmium Sulfide	Cadmium Selenide
Sensitivity	High	Very High
Dark Current	Low at Room Temp.	High at Room Temp.
Peak Spectral Response	5300–6300 Å	6800–7400 Å
Temp. Variations	Fairly Stable	Very Unstable
Breakdown Voltage	High	Medium
Life Test Stability	Good	Fairly Good
Response Time	Slow	Fast

choose cadmium selenide. Since a given application may require combinations of the characteristics of both materials, the unit may be made by mixing both materials. From the resistance needed in the illuminated state, the manufacturer determines the ratio of length to width of light-sensitive material he needs; from the power dissipation of the device he determines the area of material. A strip of photosensitive material is then printed on an insulation material, usually alumina, with the correct geometry. This process is very similar to the manufacture of thick film circuits. Two or more contacts are added to the light-sensitive material, using a compatible conducting material. Then, to protect the cell from mechanical abuse and undesirable chemical contamination, the cell is placed inside a sealed glass, glass and metal, or plastic package.

The photocell's light sensitivity occurs in the following manner. When there is no light there are very few free electrons in the semiconductor material, and the material acts almost as an insulator. When the cell is exposed to light, some of the energy is absorbed by the semiconductor material, freeing electrons. The more light energy that is absorbed, the more electrons are freed. The free electrons can move within the semiconductor material as a current. The more free electrons, the lower the resistance of the cell. Not all light energy is absorbed though. Only that energy that is in the spectral response range of the semiconductor material is effective in freeing the electrons; less than one percent for incandescent light. As we proceed through the light-sensitive

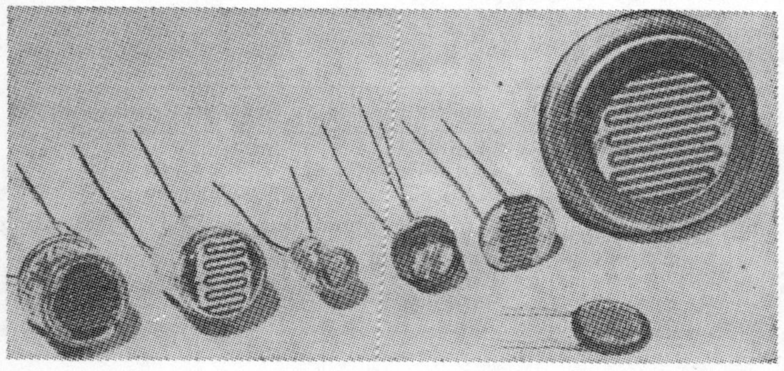

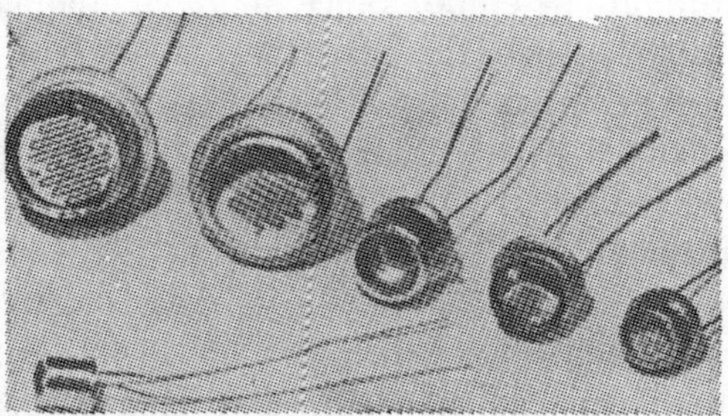

Fig. 2-2. Some common photocells. The dark lines on the face are the light-sensitive materials; the lighter background is ceramic. (Courtesy General Electric, Lighting Systems.)

devices and the light-emitting devices, it will be shown that each and every material has its own unique spectral range, and that this spectrum is based upon the allowable electron energy levels within the atoms of the material.

Photocells are easily obtainable, with most electronic component manufacturers including at least one type in their experimenters or hobby parts line for only a dollar or two. In addition, several photocell manufacturers have wide distributor coverage. The favorite packages for photocells are the standard semiconductor TO-5 and TO-8 style packages, but with glass tops, as well as a variety of glass encapsulated packages. Figure 2-2 shows some of the common packages.

When choosing photocells, several parameters are important. The voltage rating of the cell must be respected as well as its power dissipation. The light source that will illuminate the cell must be considered. The spectral response of the cell and the source must be compared to insure that the photocell can detect the light. The response times, dark leakage current, and "on" resistances should also be considered to insure proper circuit functioning.

PHOTOVOLTAIC DEVICES

Photovoltaic devices are the inverse of lamps. Lamps convert electrical energy into light energy. Photovoltaic devices convert light energy into electrical energy. The main photovoltaic device to be considered here is the photovoltaic diode or solar cell.

Available solar cells are made from either silicon or selenium. (Reports have been published on using gallium arsenide as a solar cell.) They are processed in a similar manner to any selenium or silicon diode, except that they are made of a relatively large area of silicon or selenium. The active area of these devices range from less than a square millimeter to several square inches. A 1N914 silicon diode chip is about 0.010 inches square. The present silicon cells are superior to the selenium cell in output power per unit area, stability, and the lack of memory effects. The typical power output of silicon cells is about ten milliwatts per square centimeter in sunlight. The selenium cell, however, can be made in much larger sizes than the silicon cells. The silicon cells find application in card and tape readers, position sensors, and solar power conversion. At one time, power conversion was the main use of silicon cells, but this is no longer true. At this time it costs about twenty dollars per watt of output power for silicon solar cells. The Federal Government is embarking on developmental programs to reduce this to less than five dollars per watt. The ultimate goal is in the area of one half dollar per watt, and at this price it becomes practical to consider mass electrical generation using solar conversion. The recent developments in gallium arsenide photovoltaic cells promises a chance to meet these goals.

Selenium cells find many of their applications in photographic and optical equipment, since their spectral response is most nearly equivalent to the human eye.

Fig. 2-3. Selenium photovoltaic cells manufactured by International Rectifier. Notice the large areas that are used. (Courtesy of International Rectifier.)

However, the lower cost-per-milliwatt output is increasing the use of silicon cells with filters added to compensate the spectral response. Figures 2-3 and 2-4 show some typical silicon and selenium photovoltaic cells.

The photovoltaic cells have spectral responses which are controlled by the base material. Their relative responses are approximated in Fig. 2-5. The manufacturer's data should be consulted for the exact spectral response of a given device.

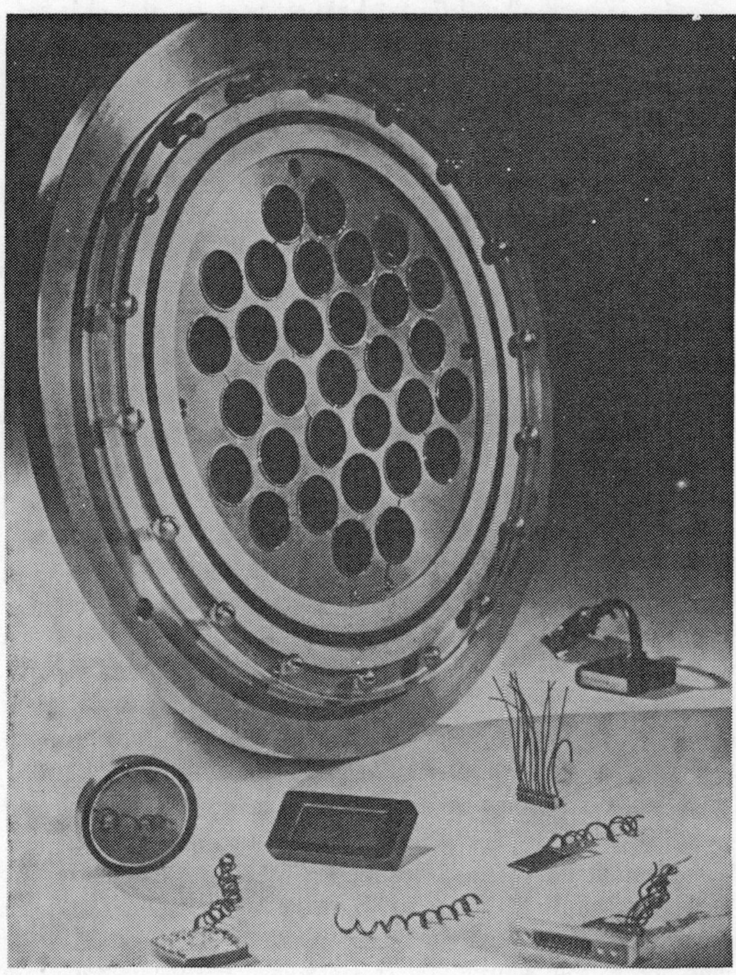

Fig. 2-4. Silicon photovoltaic cells. These cells are generally smaller than selenium cells. Several of the units shown here contain several cells rather than only one. (Courtesy International Rectifier.)

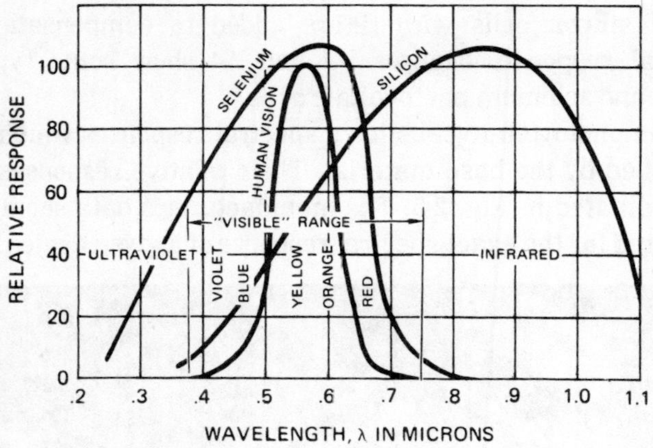

Fig. 2-5. Relative spectral response of photovoltaic materials. Notice that selenium covers the entire visual range, while silicon is basically an infrared device. (Courtesy of General Electric Semiconductor Products Dept.)

The photovoltaic cell, like the photocell, absorbs some of the energy from the illumination source. A silicon cell may only be able to use five to seven percent of the energy of the source radiation. Also like the photocells, this absorbed energy causes electrons to be freed from their atoms. The photovoltaic cell contains a pn junction which has a built-in voltage differential between the p-side and n-side, because of the doping levels. When a resistor is connected across the diode, some of the freed electrons will flow around the diode to attempt to eliminate this voltage difference. Other electrons cross the junction directly and are lost when they are reabsorbed into atoms on the p-side. As the illumination energy is increased, electrons are generated allowing more electrons to flow through the external load.

Figure 2-6 shows the equivalent circuit of a photovoltaic cell. The p and n layers represent an ideal photovoltaic cell. The electrons are freed within this area. The electrons that recombine internally in the cell are represented by the internal junction diode and its equivalent resistance, R_{sh} . R_s represents the equivalent resistance of the photovoltaic material and its leads, and R_L is the external load on the photovoltaic cell.

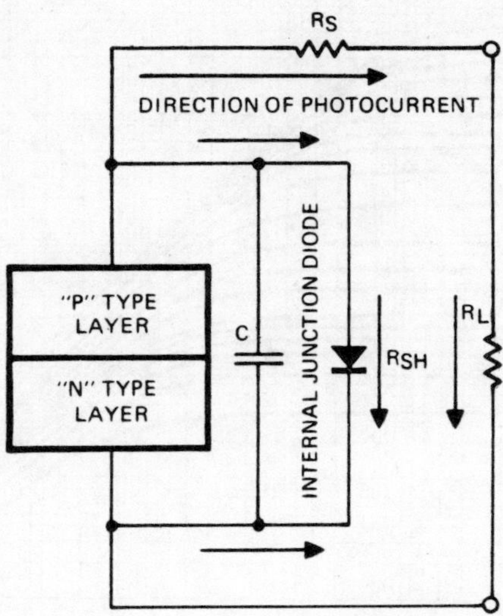

Fig. 2-6. The equivalent circuit of a photovoltaic cell. The text explains what each portion represents. (Courtesy of International Rectifier.)

The output energy of photovoltaic cells is proportional to the cell area and light intensity. Figure 2-7 shows the power output of a selenium cell as a function of the load on the cell. For high-impedance loads, the cell appears as a voltage source where the output voltage is proportional to the logarithm of the light intensity. When the cells are used with low-impedance loads, the cell appears as a current source with the current output directly proportional to the light intensity. The light intensity in many applications can be increased through the use of lenses.

THE PHOTODIODE

The most popular of the light-sensitive semiconductor devices today are the photodiode and the phototransistor. The phototransistor along with the photodarlington and photothyristor are simply amplified versions of the basic photodiode. These devices are usually specially constructed silicon devices and, therefore, have spectral responses that approximate that of the silicon curve of Fig. 2-5.

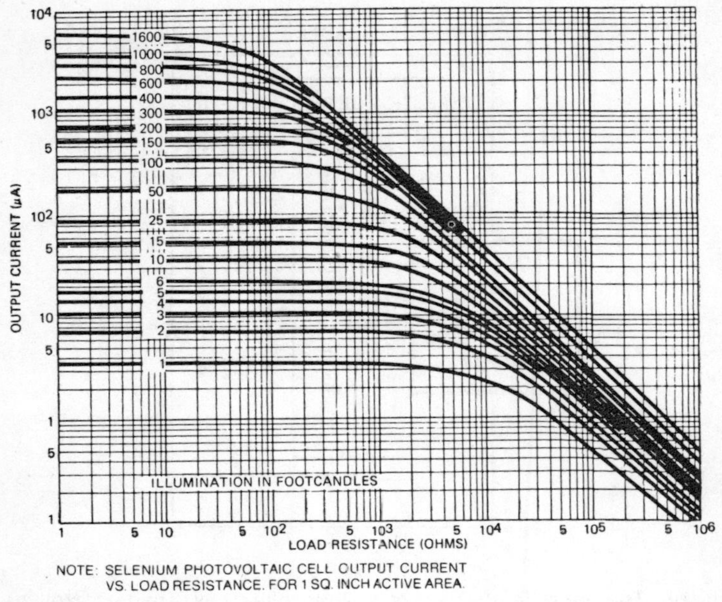

Fig. 2-7. The output of a selenium photovoltaic cell. For low resistance loads the current output changes directly with light level, while with high-resistance loads the device appears as a voltage source. (Courtesy of International Rectifier.)

To understand how the photodiode works, refer to the reverse biased pn junction shown in Fig. 2-8. With reverse bias applied to the junction, a wide region at the junction has been cleared or depleted of free charges. This region is called the depletion region. It is across this region that the electrical field exists to block the electrical current. When photons of the correct energies (i.e., of wavelengths corresponding to the spectral response of the diode material) strike the atoms in the depletion region, the absorbed energy is absorbed by the valence electrons of the atoms, freeing some of them. The free electrons leave an ionized atom behind with a net positive charge. These are called holes. Each photon with energy greater than the critical energy to free the electron will create at least one hole-electron pair when it is absorbed. The electrical field which exists in the depletion region causes these generated free charges to move, creating an electrical current. The more light that is absorbed, the more current flows. Atoms in a solid do not move, but as the electrons pass the holes some are absorbed and the net hole distribution drifts toward the negative side of the junction.

The photodiode is one of the fastest of the junction light-sensitive devices and finds applications in uses which require fast response. Examples are high-speed data transmission and laser communication systems.

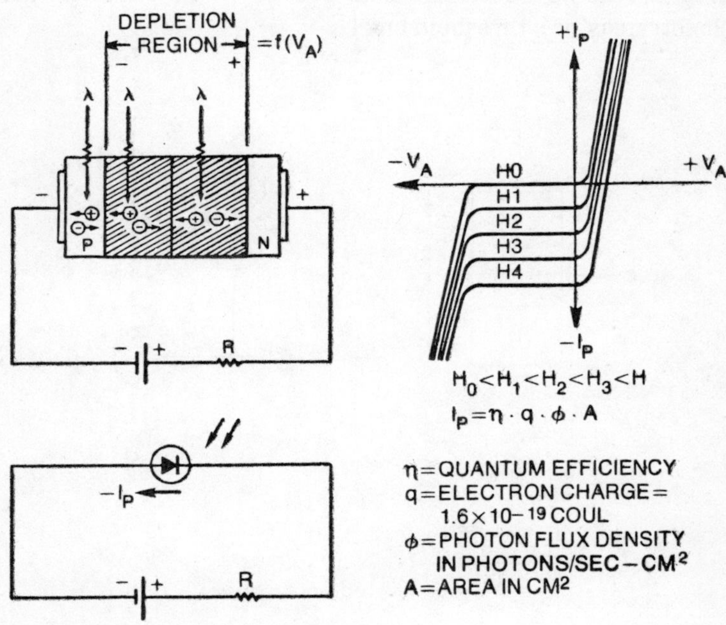

Fig. 2-8. The photodiode creates free electrons within the depletion region when exposed to light. The voltage bias on the diode pulls these electrons from the diode. (Courtesy of General Electric Semiconductor Products Dept.)

PHOTOTRANSISTORS

Photodiodes are widely used, but the photo-generated current is in the microampere range. There is a need for additional amplification in many applications. The semiconductor device manufacturers, aware of this, have provided the phototransistor to give higher light currents. In the transistor, the blocking junction is the collector-base junction;

it is this junction that supplies the photo-generated current. As shown in Fig. 2-9, the collector-base diode acts as a source of light-generated current; the base current for the phototransistor. The phototransistor's base terminal may be used to modify the collector or emitter currents. For example, by connecting a resistor from base to emitter, the photo-generated current will bypass the base-emitter junction until sufficient current flows in the resistor to cause a half-volt drop across the resistor. This results in an offset of the phototransistor's threshold level.

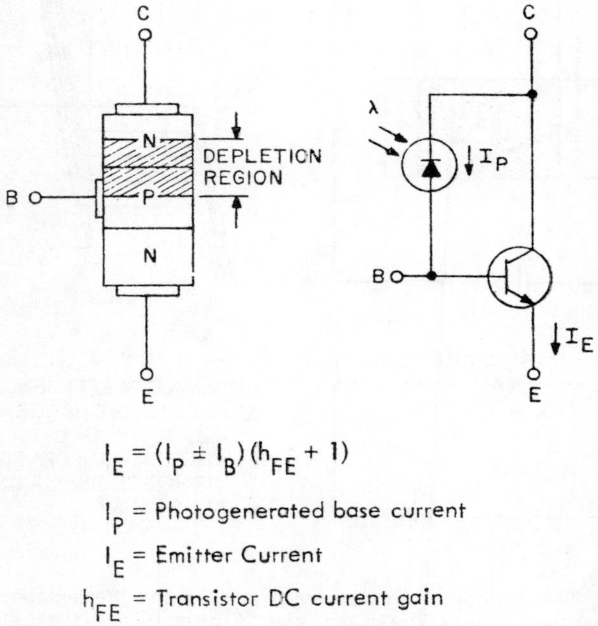

$$I_E = (I_P \pm I_B)(h_{FE} + 1)$$

I_P = Photogenerated base current

I_E = Emitter Current

h_{FE} = Transistor DC current gain

Fig. 2-9. The phototransistor's collector-base junction acts as a photodiode. The photodiode current is then amplified by transistor action. (Courtesy of General Electric Semiconductor Products Dept.)

The diode's photo-generated current is then amplified by the transistor. Thus, the phototransistor produces photo-generated currents 100 to 500 times greater than the photodiode, but the speed of the phototransistor is slowed by about the same factor. Typical switching times for phototransistors are in the one to five microsecond range.

The early phototransistors were merely standard transistors packaged in TO-5 and TO-18 packages with lens caps. Today most manufacturers have special chips designed for light-sensitive devices. If you have some metal-can transistors such as 2N2222's, the caps can be cut off to demonstrate the light sensitivity of transistors. (Some types of transistors will become contaminated within several days after removing the cap, so be prepared to discard any device which you cut open.)

PHOTODARLINGTON TRANSISTORS

The photodarlington is very similar to the phototransistor except for the additional gain provided by a second transistor. The photon current is generated now in both collector-base diodes as shown in Fig. 2-10. The photo-generated current I_{p1} has a much greater effect than I_{p2} since it is amplified by both transistor stages, whereas I_{p2} is amplified only by one transistor stage. The net effect is that the photodarlington will conduct currents of easily 10,000 times that of a photodiode. Photo-currents in the tens of milliamperes are not unusual for photodarlington transistors.

The main problem that one faces with phototransistors and photodarlingtons is speed. With photodiodes, it is possible to achieve frequency responses into the megahertz range. With phototransistors it drops to the 100 kHz range, and with photodarlingtons it is not unusual to find 5 kHz the maximum practical limit. This is due to the collector-base capacitance effects. In a typical diode, this capacitance is between 5 and 25 picofarads. The collector-base capacitance of a transistor is in the same magnitude. However, when one tries to charge or discharge the capacitance, the transistor action effectively multiplies the capacitance by the transistor's gain, thus slowing the charging and discharging rates.

LIGHT ACTIVATED THYRISTORS

SCRs, programmable unijunction transistors (PUTs), and silicon controlled switches (SCSs) are also available in light activated versions. Figure 2-11 shows the two-transistor equivalent circuit of an SCR. The blocking junction (depletion

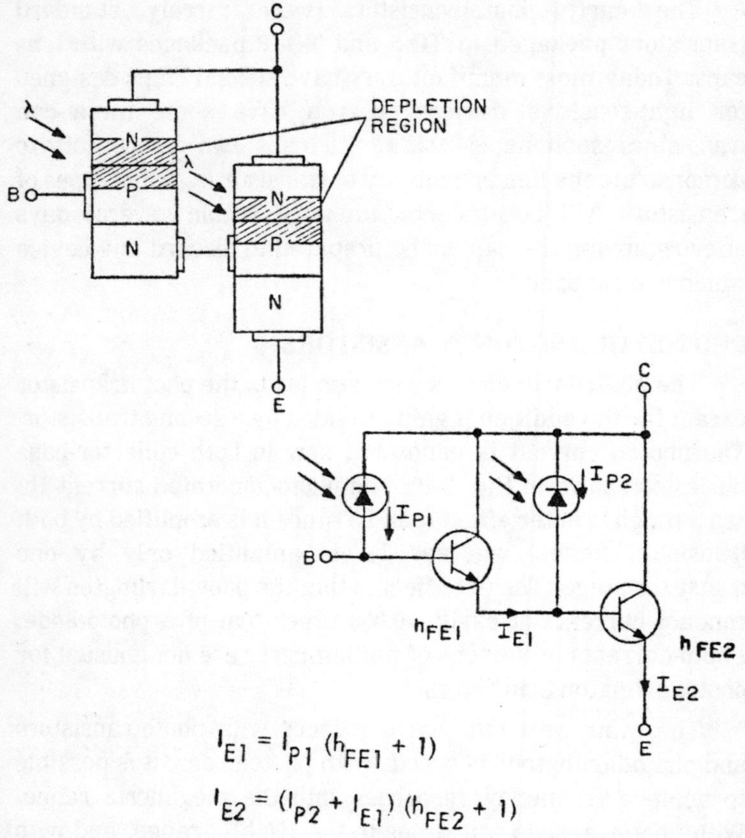

Fig. 2-10. The photodarlington transistor has two photon-generated current locations. I_{p_1} is amplified by both transistors while I_{p_2} is only amplified by one. (Courtesy of General Electric Semiconductor Products Dept.)

region) in the SCR is the collector-base junction of the equivalent transistors. When light is absorbed by the SCR, the photo-generated current flows through the emitter-base junction of each of the transistors in the equivalent circuit. The base of the npn transistor is the gate of an SCR. Hence, current through the base-emitter junction of the npn transistor appears as gate current of an SCR. Likewise, the base-emitter of the pnp transistor is the gate circuit of a PUT. It is logical that if the light current is sufficient, the thyristor can be light-triggered. In order to trigger SCRs with light, they must be constructed so that the gate current to trigger is very low, generally less than five microamperes. This gives rise to SCRs that are very sensitive to dv/dt and other dynamic effects.

At this time, there are no light-activated triacs available but it is known that at least two of the major semiconductor manufacturers are working on this device.

Light-activated transistors, darlingtons, and thyristors are all signal-level devices and can control only small loads directly. They are available in many of the small-signal transistor packages such as TO-5, TO-18, TO-92, and TO-98, which have either a lensed cap (TO-5 and TO-18) or are constructed with clear plastic (TO-92 and TO-98).

PHOTOSENSITIVE FETS

FETs are known to have a much higher frequency response than similar transistors. This is because it is not necessary to discharge the capacitance of a reverse-biased junction as in junction transistors. The FET does include a reverse-biased diode which can be used as a source of photo-generated current. This is the gate junction. Figure 2-12 shows a FET circuit. In the FET, the gate diode is operated in a reverse-biased mode. By modulating the gate bias, the depletion region expands or contracts. Since there are no free

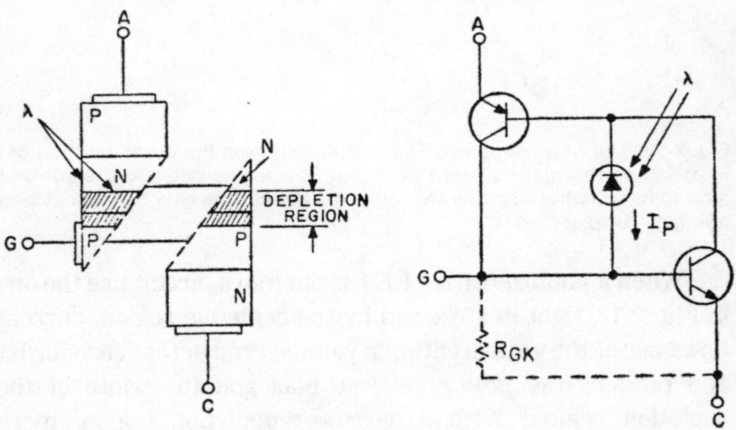

Fig. 2-11. The light-activated SCR (LASCR) has a blocking junction that acts as the light-sensitive junction. This junction supplies trigger current to the LASCR gate junction. (Courtesy of General Electric Semiconductor Products Dept.)

electrons in the depletion region, the source-drain current flows in the rest of the device. With high reverse bias on the gate the entire device becomes pinched off and only very little current flows.

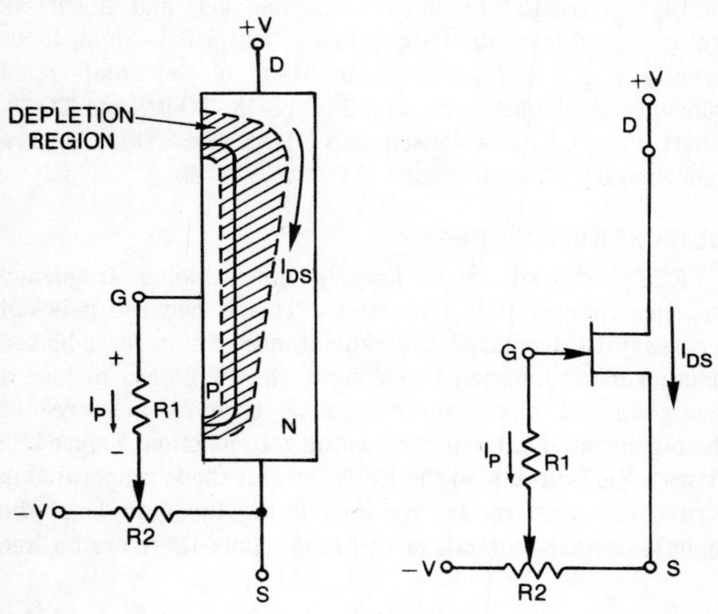

Fig. 2-12. The photosensitive FET is different from the other junction devices, since the main current does not flow in the depletion region but around it. The photon-generated current changes the gate bias by causing a voltage drop across R1.

When a photosensitive FET is put into a circuit like the one in Fig. 2-12, light is absorbed by the depletion region, current flows out of the gate creating a voltage drop across resistor R1 and reduces the gate's reverse bias and the width of the depletion region. With a narrower depletion region, more current can flow from drain to source. R2 is provided to allow the gate to be biased just at pinchoff. Resistor R1 is usually a very high impedance.

PHOTOMULTIPLIER TUBES

Among the fastest and most sensitive of the electronic photosensitive devices are photomultiplier tubes. Photomultiplier tubes are vacuum-tube devices. They contain anode and cathode electrodes plus a series of secondary electrodes called dynodes. The cathode of the tube is coated with a photoemissive material, which, when it absorbs photons, gives off electrons. A positive voltage, usually rather high, is applied at the anode. The dynodes are lined up in a precise physical structure, with each dynode at a higher potential than the preceeding one, between the cathode and anode. The electrons emitted by the cathode are drawn by the electric field to the first dynode, then the second, and so on, past ten or more dynodes. Each electron that strikes a dynode frees additional electrons. In effect, at each dynode, the number of electrons freed is multiplied.

Photomultipliers, even with their superior performance, can cause difficulties in use. In addition to the high voltage requirement (up to fifteen hundred volts) many of the tubes require cooling with dry ice or liquid nitrogen to obtain optimum performance. They are also quite a bit more fragile than semiconductors due to the vacuum-tube construction. These drawbacks are tolerated in more sophisticated applications where the very low light-level performance is a must and the higher system costs are secondary.

MISCELLANEOUS PHOTOSENSITIVE DEVICES

The semiconductor industry, with its broad range of device technologies, has generated many miscellaneous light-sensitive devices. Many of these are photodiodes which have been included within an integrated circuit. There are several digital types in which the outputs are compatible with either TTL or CMOS logic families. Several of the ICs are only available as detectors within optocouplers (see Chapter 4). There is also at least one linear IC with a photosensitive input. These devices are, in general, intended for use under a very narrow range of voltages and currents, and do not lend themselves to broad-base applications.

Another class of semiconductor devices is the photosensitive "pin" diodes. These diodes are standard diodes except that instead of direct transition from the p-type to the n-type doping regions, a thin layer of undoped or intrinsic material is left in between. Without doping, there are few free charges available, so that when the diode is reverse-biased the depletion region is wider. This lowers the junction capacitance, allowing a higher frequency response than a standard diode. The light current is generated in an identical manner to a standard pn diode.

Chapter 3

Light-Emitting Devices

In the previous chapter it was shown that there are many varieties of devices which are sensitive to light. There are also many sources of light, ranging from the sun, fire, and some natural materials, to the lasers and other devices which have been developed within the past decade or so. In this chapter, only the electrically activated light sources will be considered.

Light is generated by many sources, but all of these sources can be divided into two classes; incandescent and illuminescent. The first includes all those sources where the light emitted is due to the high temperature of the source. The tungsten lamp is a common example of this type of light source. The filament is heated to a temperature of from 2100°C to 2800°C, and because of its temperature it emits radiation which we perceive as light. In actuality, the filament emits radiation over a very broad continuous spectrum, but we see only the visual portion of that spectrum.

The illuminescent sources, on the other hand, emit radiation at only certain defined wavelengths. Gas discharge lamps, such as neon glow lamps, are important light sources of this class. The reddish color of the neon glow lamps is due to the characteristics of the neon atom, whereas sodium vapor lamps yield a yellow color.

INCANDESCENT LAMPS

On an October day in 1879, Thomas A. Edison demonstrated the first practical incandescent lamp at Menlo Park, New Jersey. This first lamp used a filament made of sewing thread which had been burned to leave a bar of carbon. When an electric current was passed through the carbon, it heated the carbon until it was hot enough to glow and give off light. The following year a lamp was being manufactured and sold for $1.00; each used a carbonized-bamboo filament.

Since the carbon filament was relatively fragile, there was a lot of breakage in shipment and use. This caused a search for a new and better filament material, and for years many materials were tried and several were manufactured and sold. These were made from osmium and tantalum among other materials. In 1908, the world was introduced to the tungsten filament which is still used today. As the years passed, many advances were made in incandescent lamp design. Today, we have lamps which are over ten times as efficient as the first carbon filament lamps.

PHYSICAL CHARACTERISTICS

To many of us an incandescent lamp is just a glass bulb with a filament inside and a base attached, but in reality there are other parts which are just as important. Figure 3-1 shows these parts for a standard incandescent lamp.

The filament, as mentioned, is usually a length of tunsten wire coil which is heated to incandescence by electric current. Coiling the first coil (called a coiled-coil filament) further increases its efficiency. The gauge of the wire and its length determine its resistance and therefore the voltage and current needed. An automobile headlamp has a heavier wire than an equivalent wattage 120-volt lamp since, at 12 volts, ten times the current is required for the same wattage.

The filling gas is usually an inert mixture of nitrogen and argon, but for small lamps the glass bulb may contain a vacuum. The purpose of the gas is to reduce the evaporation of the tungsten filament and therefore allow higher filament temperatures. The higher temperature produces greater lamp efficiency.

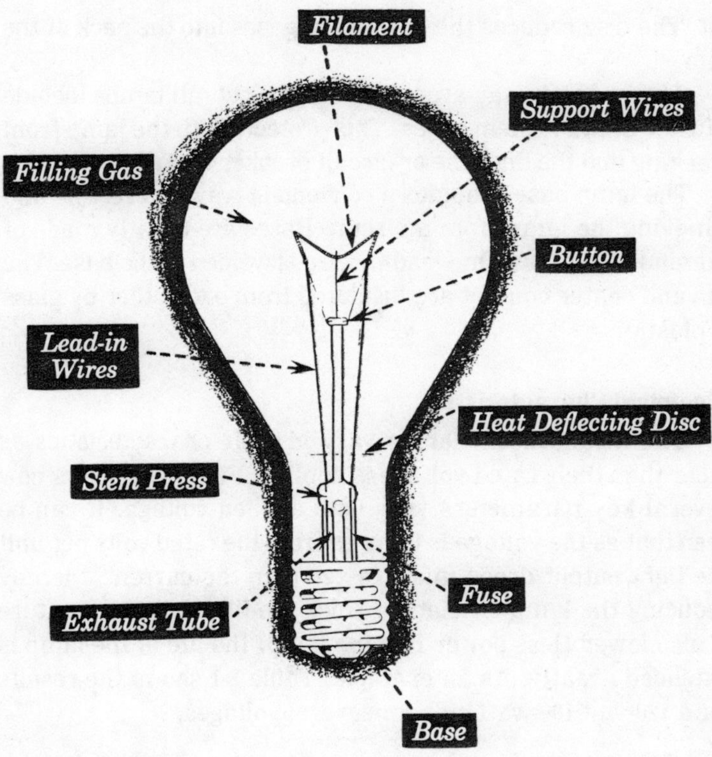

Fig. 3-1. The parts of an incandescent lamp. The incandescent lamp has come a long way since Thomas Edison. (Courtesy of General Electric Lamp Division.)

The lead-in wires electrically connect the filament to the lamp base through the stem press. The stem press is made of glass. The wires within the stem press are made of a special combination of a nickel-iron alloy core with a copper sleeve. This wire, called Dumet wire, has substantially the same coefficient of expansion as the glass and, thereby, allows a good glass-to-metal seal to maintain the vacuum or filling gas. The button rod protrudes from the top of the stem press. The button atop the rod secures two filament support wires. The support wires hold the filament in place and protect it during shipment and in use.

For the higher-wattage bulbs a heat deflecting disc is added to prevent the stem press and base from becoming too hot. The disc reduces the flow of hot gases into the neck of the bulb.

Since a lamp may arc when it burns out, all lamps include a fuse within the stem press. This protects both the lamp from cracking and the line fuse or circuit breaker from opening.

The lamp base provides a convenient way of inserting and removing the lamp from a circuit. Bases are usually made of aluminum or brass. One lead-in wire is welded to the base. The rim and center contact are insulated from each other by glass or plastic.

Electrical Characteristics

The incandescent lamps vary in their characteristics as other than their rated voltage is applied. Figure 3-2 shows how several key parameters vary with applied voltage. It can be seen that as the voltage is lowered from the rated volts per unit the light output drops much faster than the current, thereby reducing the lamp efficiency. Since the filament temperature is also lower (less power into the lamp) the life of the lamp is extended greatly. As an example, Table 3-1 shows the results for a 120-volt 100-watt bulb for several voltages.

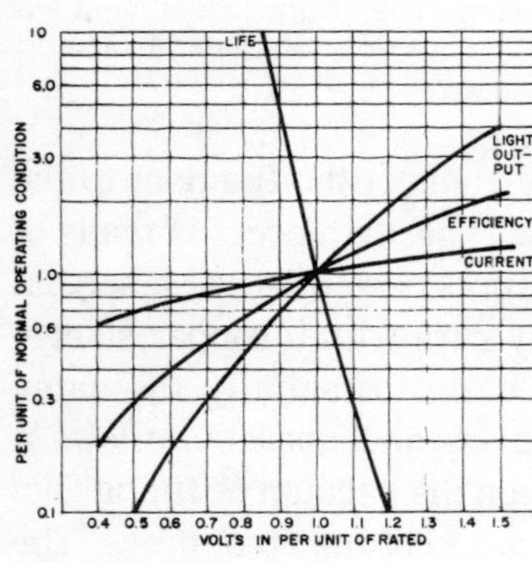

Fig. 3-2. The incandescent lamp is not a constant resistance with changes in voltage. The light output and life change drastically with relatively small changes in voltage. (Courtesy of General Electric Semiconductor Products Dept.)

Due to the characteristics of tungsten wire, incandescent lamps can provide unexpected problems to those who try to use them in electronic circuits. In Table 3-1 it can be seen that the current does not vary linearly with voltage. If the current

Table 3-1. Changing the Voltage on an Incandescent Lamp Can Drastically Change Its Characteristics

% RATED VOLTAGE	VOLTAGE	POWER IN WATTS	LIGHT OUTPUT IN LUMENS	LIFE IN HOURS
120	144	133	7900	80
110	132	116	2320	280
100	120	100	1680	1000
90	108	85	1170	4000
80	96	71	790	18500

was linear, then at 80% of rated voltage the current would be 80% of rated current and therefore 64 watts rather than the 71 watts. This trend continues all the way to zero voltage. The result is that a cold lamp will have a resistance of one-fourteenth to one-twentieth of its resistance with rated voltage applied. This is due to the temperature coefficient of resistance of tungsten. To the unwary, this can create many unexpected problems. Since the cold lamp has a much lower resistance, it will conduct very large currents until it heats up to temperature. This requires that the semiconductors used to drive lamps have sufficient current capability to carry this inrush of current. As an example of this inrush of current, Fig. 3-3 shows a 1500-watt projection lamp inrush.

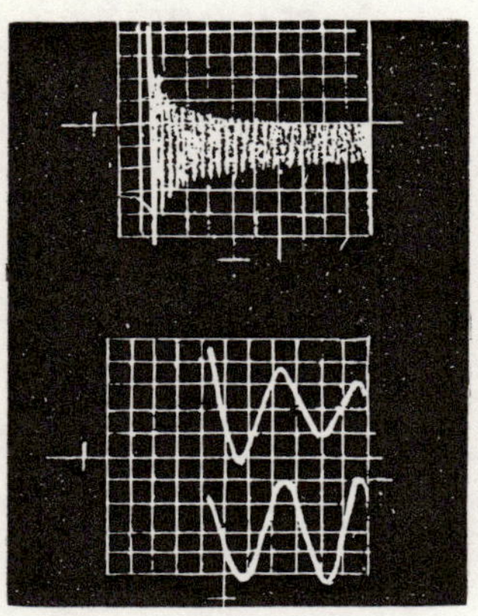

(a) Decay of Inrush Current
(Scale: 40 A/Division)

(b) Top: Inrush Current
Bottom: Voltage Showing Switch Closure at Approximately 85 Electrical Degrees
(Current Scale: 100 A/Division

Fig. 3-3. These photographs of oscilloscope traces show the extremely high currents which can be encountered during incandescent lamp turn-on. The 1500-watt lamp of these traces has a normal current of 12.6 amperes, while the tracer shows peak currents of over 250 amperes. (Courtesy of General Electric Semiconductor Products Dept.)

ILLUMINESCENT LAMPS

Nature provided the first light source from an electrical discharge through a gas—lightning. When the electrical potential between clouds, or between clouds and the earth, exceeds the insulating capability of the air between, an electrical discharge occurs, creating a lightning bolt. The electrical energy ionizes the air, creating an unstable condition. As the air returns to its normal state, photons are given off which we see as lightning. Of course many other factors are involved in this process, but this is similar to the phenomena which occur in gas discharge lamps.

Since each element has its own unique combination of electrons, it is logical to assume that each has its own set of unstable or excited states, and in fact each one does. The electrons in the gas are put into the higher energy states by either radiation absorption such as heat, or by collisions with fast moving electrons. The former method is somewhat impractical to accomplish electrically but accounts for much of the light given off by carbon arc lamps and the like. Moving electrons is no problem at all in electronics since we do it all the time. In fact, the vacuum tubes which we all have used

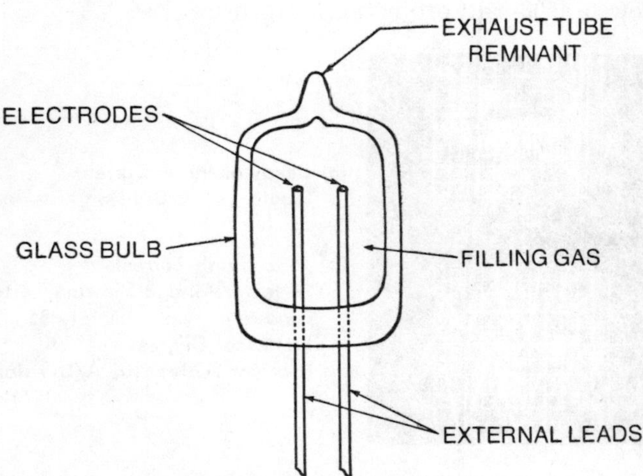

Fig. 3-4. The neon glow lamp's structure is much simpler than the incandescent lamp of Fig. 3-1. This makes the neon glow lamp one of the least expensive lamps available.

operate in a similar manner to gas discharge lamps, even to the use of filaments as sources of electrons.

Neon Glow Lamps

The simplest form of gas discharge lamp is the neon glow lamp. As shown in Fig. 3-4, it consists of only a glass bulb with two electrodes, which is then filled with neon and sealed. Since the lamp is constructed symmetrically, that is, both electrodes are the same, the electrical characteristics are also symmetrical. The typical characteristics of a neon lamp are shown in Fig. 3-5. When a neon lamp such as this is used on a normal 120 VAC line, some impedance must be used to limit the current through the lamp. For most normal applications a resistor is used for this purpose.

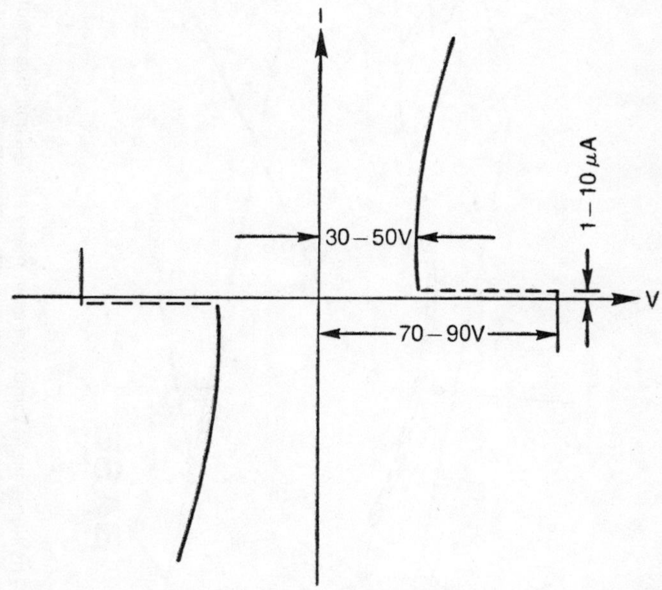

Fig. 3-5. The neon glow lamp has a non-linear characteristic. The lamp does not turn on until its breakdown voltage is exceeded. A series resistor will keep the lamp current within limits.

Fluorescent Lamps

The neon glow lamp is an example of the cold-cathode lamp since there are no heating elements for the two

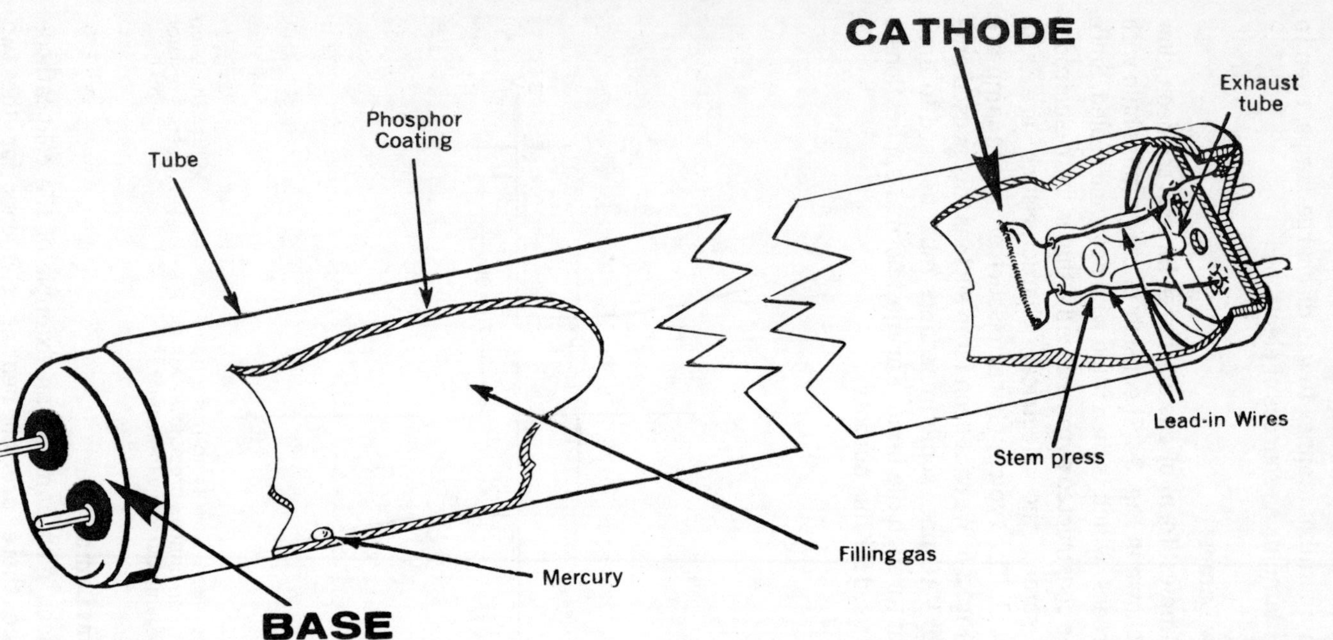

Fig. 3-6. The fluorescent lamp comes in many shapes and sizes. Notice how the cathode structure is very similar to an incandescent lamp or vacuum-tube filament. (Courtesy of General Electric, Lamp Division.)

electrodes or cathodes. An example of a heated-cathode lamp is the normal fluorescent lamp. Figure 3-6 shows the basic parts of the fluorescent lamp. It can be seen that the cathodes are essentially the same as the internal parts of the incandescent lamp (Fig. 3-1). The main differences are the lower operating temperature of the tungsten filaments and the filling of the filament with alkaline-earth oxides to improve discharge lamp efficiency. There is one cathode at each end of the tube with each end of each filament connected to one terminal of the base, for a total of four terminals; two at each end.

The tube acts as an airtight enclosure for the cathode, the phosphor coating, and the filling gas. Mercury, the filling gas, is vaporized at a very low pressure (about 1/100,000th of atmosphere pressure). The phosphor coating is used to change the wavelength of the mercury radiation (253.7 nonometers, which is in the ultravoilet region) into visible light. The phosphor coating is made up of small particles, approximately 0.0007-inch in diameter.

The filaments of the fluorescent lamp are heated to supply a source of electrons, and a high voltage is applied between the two cathodes, accelerating the electrons down the tube. As these electrons travel down the tube, they strike mercury atoms, creating photons which are absorbed by the phosphor coating on the inside of the tube. The phosphors then give off other photons which are the light we see.

By changing the phosphor, different colors can be achieved. Once the lamp starts, a finite number of electrons exist along the tube, changing the lamp's characteristic much like in the neon glow lamp. This again gives need for some type of ballasting arrangement. A resistor ballast has proved effective only for the smallest of fluorescent lamps, generally those of eight watts and below.

Most fluorescent lamps are divided into three categories: preheat lamps, instant start lamps, and rapid start lamps. Each of these types require their own unique ballast system and cannot be interchanged in fixtures without also changing the ballast.

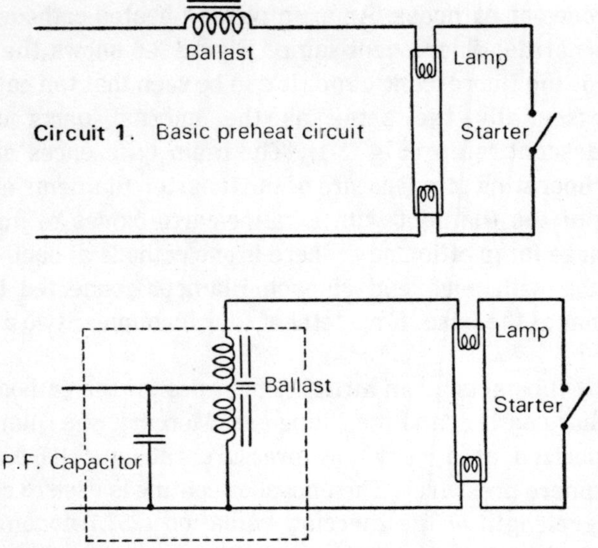

Circuit 1. Basic preheat circuit

Circuit 2. Preheat circuit with autotransformer to step up voltage and capacitor to correct power factor

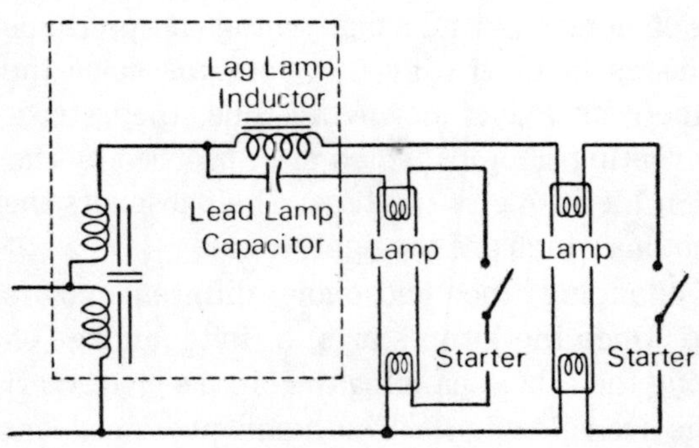

Circuit 3. Lead-lag preheat circuit

Fig. 3-7. Preheat fluorescent lamp circuits. Circuit No. 1 uses only a choke to limit lamp current. Two other representative preheat circuits (No. 2 and 3) are shown; these circuits are autotransformers to limit current and increase voltage. The capacitor across the line in Circuit No. 2 corrects the power factor. In Circuit No. 3, a lead-lag system, the lamp circuit controlled by the capacitor has a leading power factor while the lamp circuit controlled by the inductor has a lagging power factor. Together the two circuits result in essentially unity power factor. (Courtesy of General Electric, Lamp Division.)

Early fluorescent systems were all of the preheat type, requiring separate starters (See Fig. 3-7). A few seconds were required between the time the current to the lamp was turned on and the lamp lighted. During these seconds the filaments were heated to provide electrons to start the lamp. Once the lamp was on, the lamp current provided sufficient heat to keep the cathodes emitting electrons and the heaters were no longer required. An automatic starter was used to switch the filaments in and out of the circuit. Figure 3-8 shows the three main types. A preheat system can immediately be recognized by its starter. (Desk lamps are usually preheat systems with a manual starter in the off-on switch.)

Instant-start lamps are started by applying an extra high voltage to start the lamp by voltage breakdown. The lamp current heats the very fine cathode wires to provide the electron source. Since there is no need for an external heater source, most instant-start lamps have only a single pin at each end. The lack of preheating circuitry also simplifies the ballast and the fixture wiring. Figure 3-9 shows three of the basic instant-start circuits.

The rapid-start system is a combination of the preheat and the instant-start systems. Like the preheat system, the rapid-start system has heated filaments but they are designed for continual energization and consequently have no starter. They are started with a high voltage, but because of the heated filaments they do not require as high a voltage as the instant starts, giving a reduction of ballast cost and an increase in system efficiency. Figure 3-10 shows the common rapid-start circuits. Because of the continuously heated filaments, these lamps are the type used for dimming and flashing applications, as shown in Fig. 3-11.

The fluorescent, neon glow, and sign lamps are referred to as low-pressure lamps, since the vapor pressure of the gases is less than atmospheric pressure.

Mercury Vapor Lamps

There are other lamps which are referred to as high-pressure lamps, available in both the cold-cathode and preheat types. An example of this group is the mercury vapor lamp. The cold-cathode mercury vapor lamps usually contain

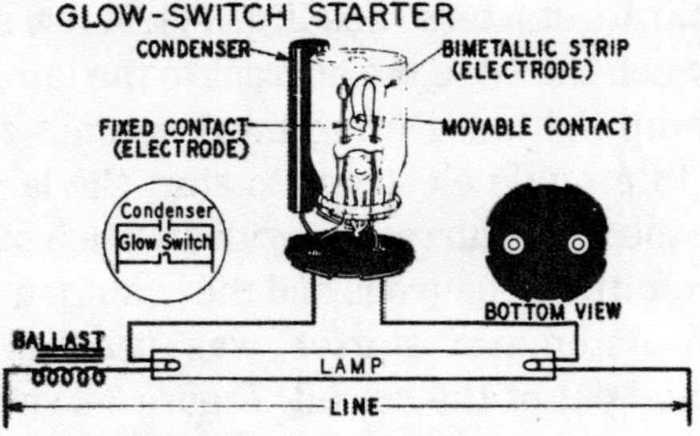

GLOW-SWITCH STARTERS. On starting, voltage at the starter produces a glow discharge. Heat from the glow discharge actuates the bimetallic strip, the contacts close, extinguishing the glow, and cathode preheating begins. When the bimetal cools sufficiently due to the elimination of the glow, the contacts open, and the lamp starts. During operation, voltage across the starter is too low to produce further glow; the contacts remain open.

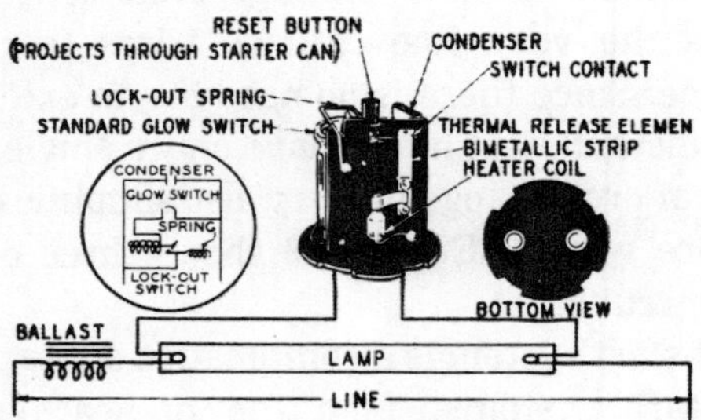

WATCH-DOG* STARTERS. This type of starter automatically removes the glow switch from the circuit if the lamp fails to start after 15 or 20 seconds. This eliminates annoying lamp blinking, and conserves the life of the ballast. A new metal glow switch in the GE FS-400 Watch-Dog starter reduces instant-starting tendencies of the 40-watt lamp on lead circuits. This tends to increase lamp life.

*Registered T.M.

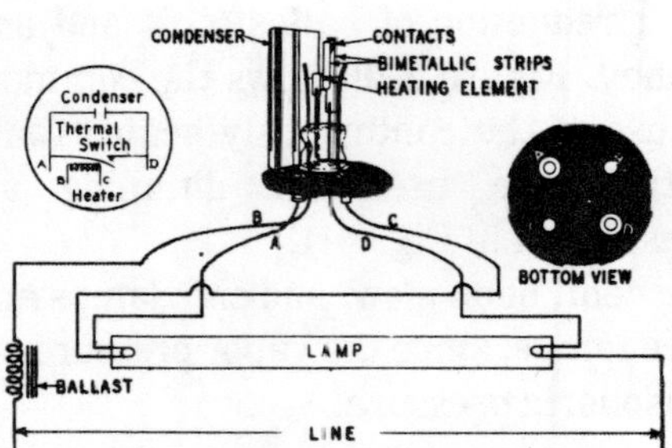

THERMAL-SWITCH STARTERS. This type of starter is useful in operating conditions involving low temperature, direct current, or widely varying line voltage. It has a heating coil in series with the ballast and lamp. When voltage is first applied to the circuit, preheating occurs immediately since the thermal switch is closed. The heating coil activates the bimetallic strip in the starter switch. After sufficient preheat time elapses, the thermal switch opens, and the lamp starts. During lamp operation a small amount of energy is consumed by thermal starters.

Fig. 3-8. The three main types of starters used for preheat fluorescent lamps. (Courtesy of General Electric, Lamp Division)

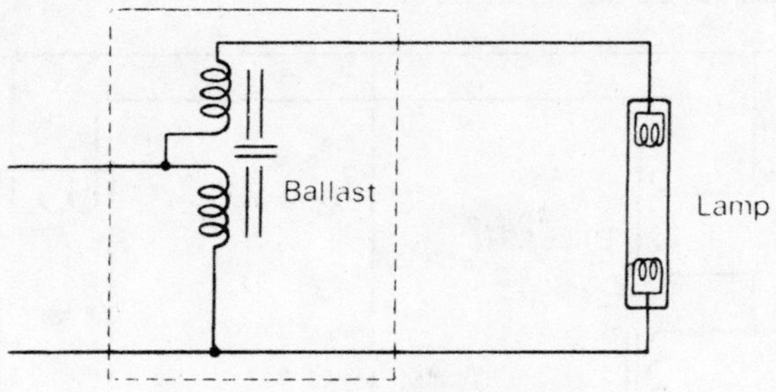

Circuit 1. Basic instant start circuit

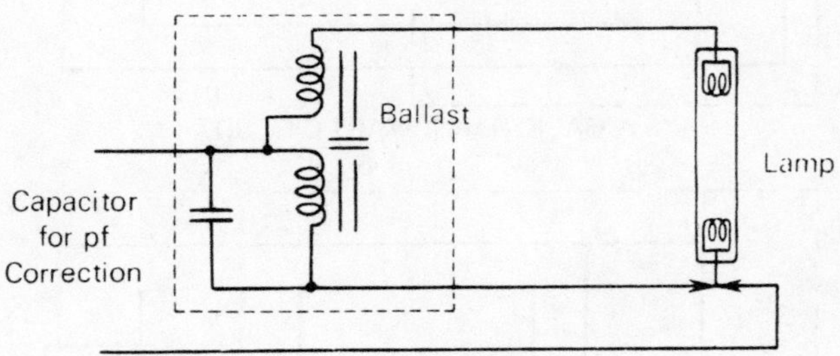

Circuit 2 Instant start circuit showing disconnect lampholder

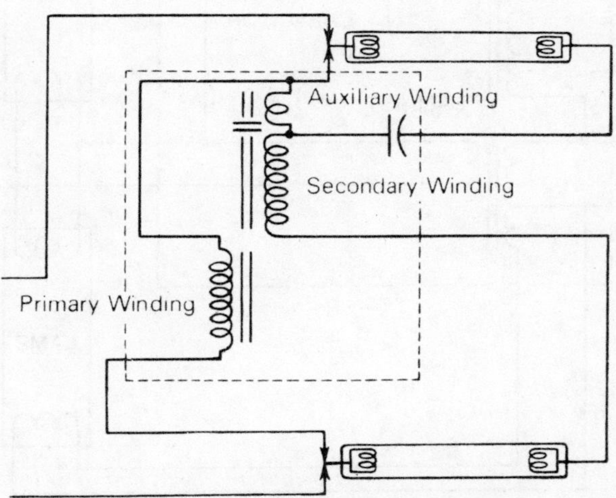

Circuit 3. Typical series instant start circuit

Fig. 3-9. Three of the instant-start ballasting circuits. Notice that there are no starters used. Instead the initial voltage from the ballast is much higher and the lamp is started by a voltage breakover. (Courtesy of General Electric, Lamp Division)

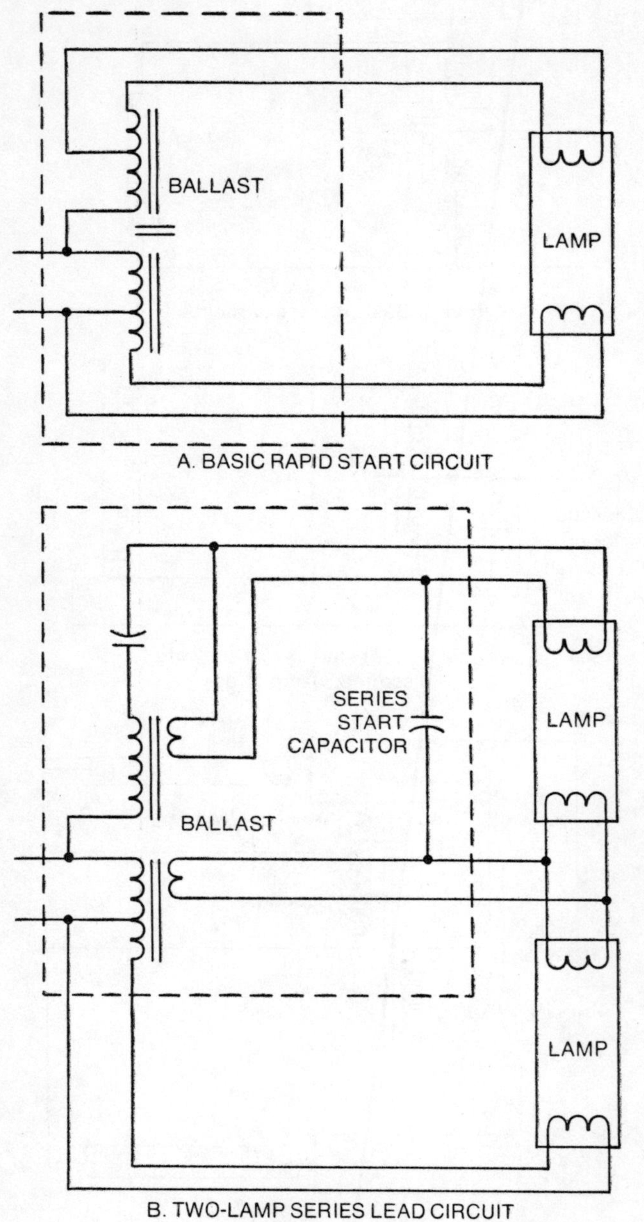

Fig. 3-10. The rapid-start ballasting arrangement is a compromise between the preheat circuitry with the heater windings and the instant-start which uses no starter. (Courtesy of General Electric, Lamp Division)

neon at a low pressure and a pool of mercury. Starting is accomplished by subjecting the lamp to a high enough voltage to start the lamp as a neon lamp. The lamp current then heats and vaporizes the mercury. The preheat mercury vapor lamps use a filament to vaporize the mercury. In the screw-base mercury vapor lamps, the filament is directly across the arc electrodes so that it appears to be a filament lamp when viewed. The addition of tungsten electrodes on the lead-in wires provide the cathodes for the discharge current.

LED Theory

In gas discharge lamps, the high-energy free electrons stike atoms and gives up some of their energy to the atoms, creating photons. In light-emitting diodes (LEDs) or solid-

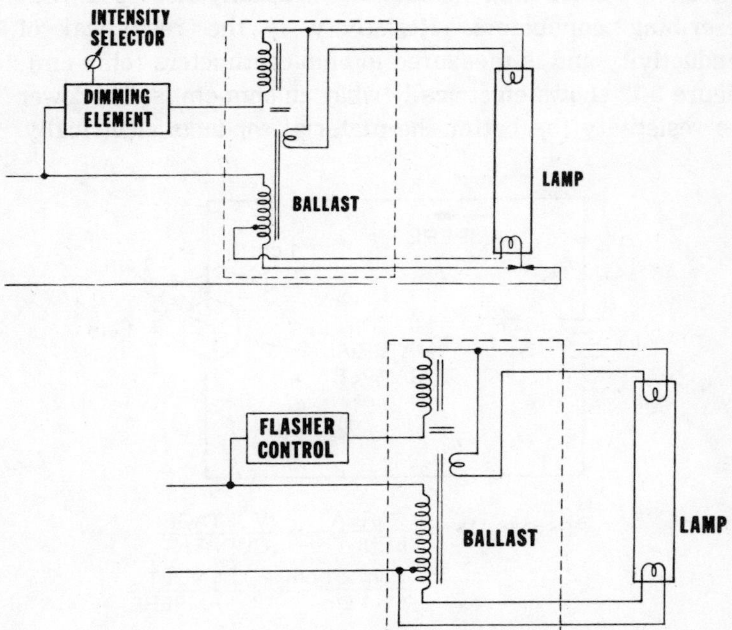

Fig. 3-11. The circuits for fluorescent dimming and flashing applications. Special ballasts are available for these applications. In each of these the cathodes are continuously heated while the control circuitry only controls the high voltage to the lamp. (Courtesy of General Electric, Lamp Division)

state lamps a similar phenomenon occurs. All LEDs are made from semiconductor materials. In order to understand how light is generated in these devices some basic semiconductor theory should be discussed.

The electrical characteristics of materials are divided into three classes: conductors, insulators, and semiconductors. Materials such as metals are good conductors of electricity, so they're called conductors. Glass and ceramics are very poor conductors and are termed insulators. In between conductors and insulators there is a third class called semiconductors. Many materials fall into this class and they all have one property in common: they are poor conductors of electricity.

The ability of a material to conduct electricity is a function of the number of free electrons it contains. The more free electrons the higher its conductivity. The metals have many free electrons and have a very high conductivity. However, resistivity rather than conductivity is usually discussed when describing conductors. Resistivity is the reciprocal of conductivity and is measured in ohm-centimeters (ohm-cm). Figure 3-12 shows electrically what an ohm-cm is. The lower the resistivity the better the material conducts electrically.

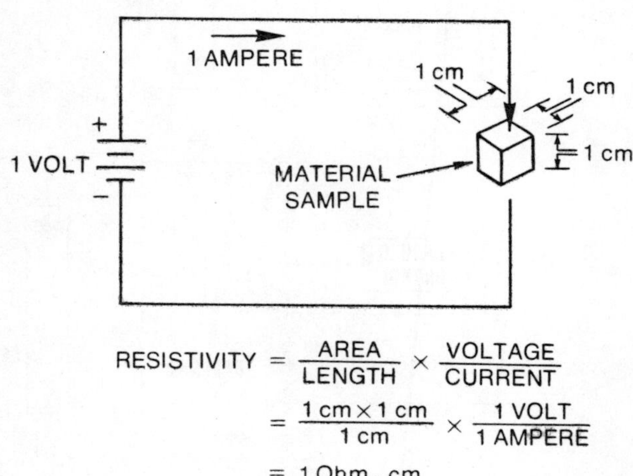

$$\text{RESISTIVITY} = \frac{\text{AREA}}{\text{LENGTH}} \times \frac{\text{VOLTAGE}}{\text{CURRENT}}$$

$$= \frac{1\text{ cm} \times 1\text{ cm}}{1\text{ cm}} \times \frac{1\text{ VOLT}}{1\text{ AMPERE}}$$

$$= 1\text{ Ohm-cm}$$

Fig. 3-12. The resistivity of a material tells us how well it conducts electric current. If the sample was two inches long and still conducted one ampere, then the resistivity of the material would be 0.5 ohm-cm.

Conductors such as metals have resistivities of much less than 0.01 ohm-cm, where insulators have resistivities greater than 100,000 ohm-cm. Between these extremes lie the semiconductors. Figure 3-13 shows this relationship.

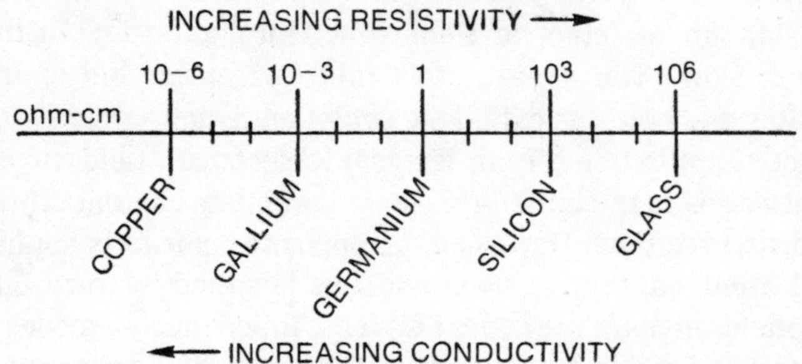

Fig. 3-13. Materials range widely in their ability to conduct electricity. On one hand, metals like copper are good conductors—low resistivity. On the other hand are poor conductors, which we call insulators, like glass. In between, lie the broad spectrum of semiconductors.

When LEDs are made, the manufacturers start with a pure semiconductor material. Depending upon the type of material it is either a single crystal or polycrystalline (many crystals) material. LEDs are made from gallium arsenide, gallium phosphide, or a mixture of the two, depending on the color or wavelength of light output desired. Small amounts of impurities called dopants are added to the pure material. Some dopants create excess electrons within the material, while others give the material the ability to absorb electrons. Each atom that can absorb an electron is called a hole. Holes can be considered a positive charge. Therefore, if an atom loses an electron, the formerly uncharged atom creates a hole and an electron, resulting in two free charges but with a net charge of zero.

In manufacturing different dopants are used so that different parts of the device either have holes or electrons. The portion of the material with excess holes is called p-type; the material with excess electrons, n-type. (P-type for excess positive charge, n-type for excess negative charge.) When n-type and p-type are adjacent, we have a pn junction where they meet.

At the junction a few electrons are drawn across by the adjacent positive charge (holes). The electrons recombine with the holes, producing a slight negative charge in the p-type material and a slight positive charge in the n-type material. This sets up an electric field which prohibits any further electron flow. The area of this field, which is void of free electrons and holes, is called the depletion region.

The application of an external electrical field in one polarity adds to the existing field and prohibits current. This is called reverse bias. But when the opposite polarity is applied, it will allow current to flow, and this is called forward bias. This phenomenon is identical for all semiconductor diodes or pn junction devices. But to understand how light is produced we must investigate an additional portion of semiconductor physics.

According to our modern laws of physics, an electron in orbit around an atom is only permitted certain discrete energy levels. Each of these energy levels corresponds to a discrete orbit. Energy levels and orbits other than these are forbidden. This holds only for a free, single atom, such as in a low-pressure gas. When many atoms are crowded together in a solid, these levels get smeared out or broadened into bands. These bands are centered around each of the allowed levels just discussed. Between bands there are still forbidden bands.

Electrons tend to seek the lowest possible energy level, so they tend to fill the bands from the bottom up. In pure semiconductors there are two bands of particular interest. They are the valence band and the conduction band. These two bands are separated by a forbidden band in which no electrons exist. The valence band is the resting place of the valence electrons, which take part in making chemical compounds. The conduction band contains the free electrons which are measured by the conductivity or resistivity of the material as mentioned previously. Therefore electrical current can only exist when there are sufficient electrons in the conduction band. In semiconductors, at room temperature, there is a finite number of electrons available, but as the temperature is lowered, these electrons lose their excess energy and fall back to the valence band and the material becomes an insulator.

Doping the semiconductor n-type creates an excess of electrons in the conduction band; doping the semiconductor p-type creates an excess of holes in the valence band. This is shown in Fig. 3-14.

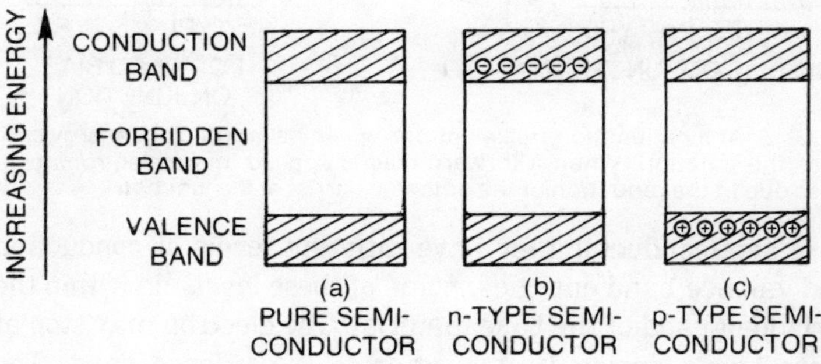

Fig. 3-14. In a pure semiconductor (a) there are few free, mobile charges. In an n-type (b) there is an excess of free electrons in the conduction band. In a p-type (c) there is a lack of electrons in the valence band, which we call holes. Holes are regarded as positive charges.

When a pn-junction is formed, as discussed earlier, the potential barrier prevents the conduction band electrons from moving into the p-type material as shown in Fig. 3-15a. If the junction is forward biased (Fig. 3-15b) the barrier is lowered and the mobile electrons move across the junctions. Note that only those electrons with energies greater than the now reduced barrier move. There now exists an excess of electrons in the p-type material which is in an unstable state. These electrons quickly recombine with the holes in the p-type valence band in order to restore the equilibrium.

The recombination results in a decrease in energy of the electron. This energy may be given off in the form of heat, light, or both. There are four main modes of recombination as shown in Fig. 3-16. The simplest is shown in Fig. 3-16a where the electron recombines directly with the hole. The other three modes of recombination involve the characteristics of the doping atoms. Since they are of a different material than the

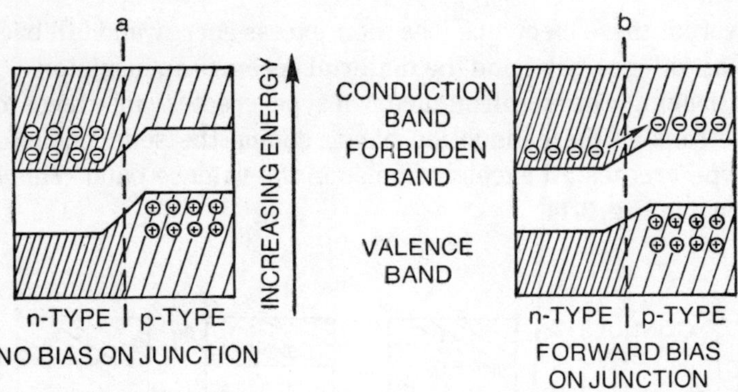

Fig. 3-15. At a pn junction,free electrons which have sufficient energy can cross the junction. When a forward bias is applied, more electrons can cross due to the reduction of the potential barrier at the junction.

base semiconductor they have different levels of conduction and valence band energies. Some of these levels lie within the forbidden band of the base material. The electron may stop at these levels during its transition to the valence band. The levels in the upper or higher energy portion of the forbidden band are due to the donor atoms, while those near the bottom are due to acceptor atoms. These later transitions, as shown in Fig. 3-16b, c, and d, result in heat as well as light.

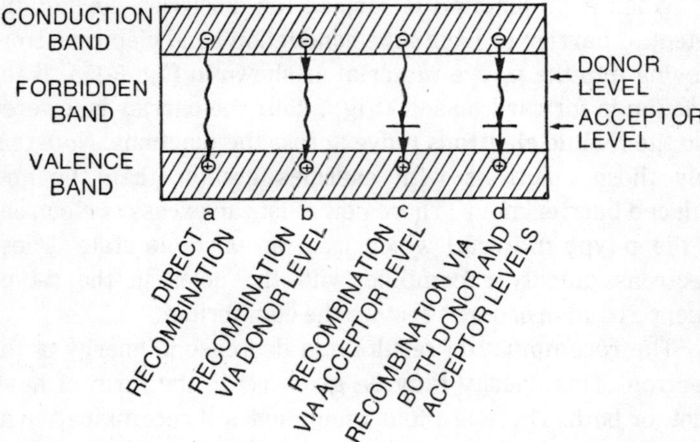

Fig. 3-16. Electrons in the conduction band will recombine with available holes in the valence band. The electrons may recombine directly (a) or recombine after a stop at an empty donar or acceptor level (b and c) or both (d). The wavy lines represent photons given off. The straight lines represent heat given off.

Now that we have seen why and how LEDs emit light, we can understand that the electrical characteristic of an LED is the same as any pn-junction device. It conducts current when forward biased and blocks voltage when reverse voltage is applied. (LEDs are not known for good reverse characteristics and many can be damaged with more than a few volts reverse bias.)

The LED, being a semiconductor device, is packaged as many other devices are: either encapsulated in plastic or placed within a hermetic package. The only difference is that for LEDs the package must let the light out so that it can do something useful. Figure 3-17 shows a typical hermetically packaged LED.

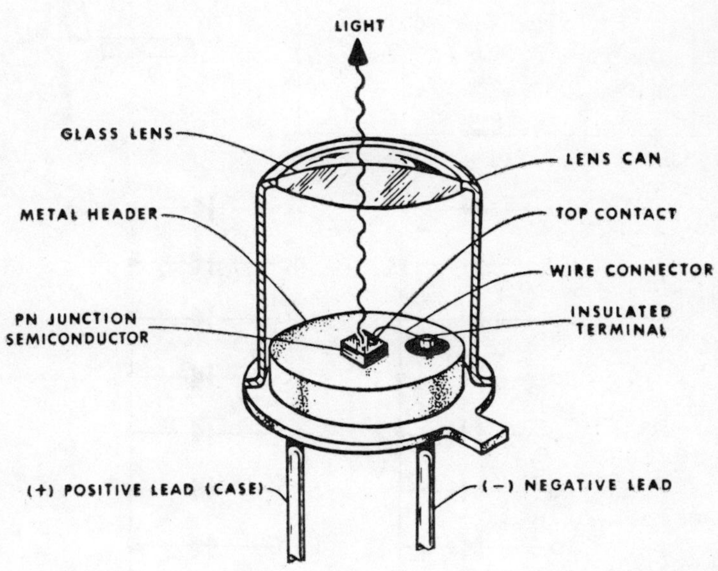

Fig. 3-17. This is a typical package of an LED. The lens allows the photons to leave the package. (Courtesy of General Electric Semiconductor Products Dept.)

LED Displays

In operation, LEDs typically operate with forward currents of 10 to 25 milliamperes and a voltage drop of 1.5

volts. This makes them extremely valuable in giving visual signals from minicomputers and other TTL and MOS integrated circuits. Aware of this, manufacturers have created special devices such as the seven- and eleven-segment readouts to permit more information to be transmitted optically from integrated circuits.

CHARACTER	SEGMENTS	DISPLAY
0	a, b, c, d, e, f	
1	a, b	
2	a, b, d, e, g	
3	a, b, c, d, g	
4	b, c, f, g	
5	a, c, d, f, g	
6	(a), c, d, e, f, g	
7	a, b, c	
8	all	
9	a, b, c, f, g	

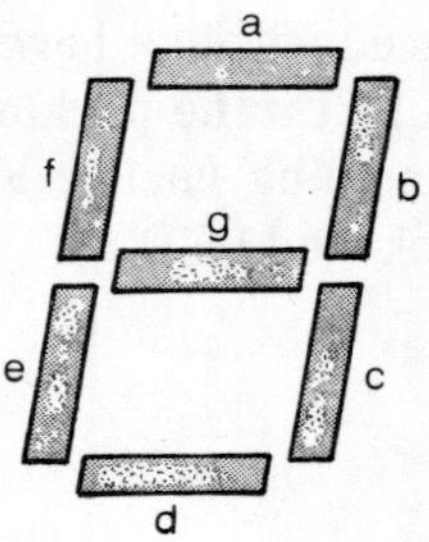

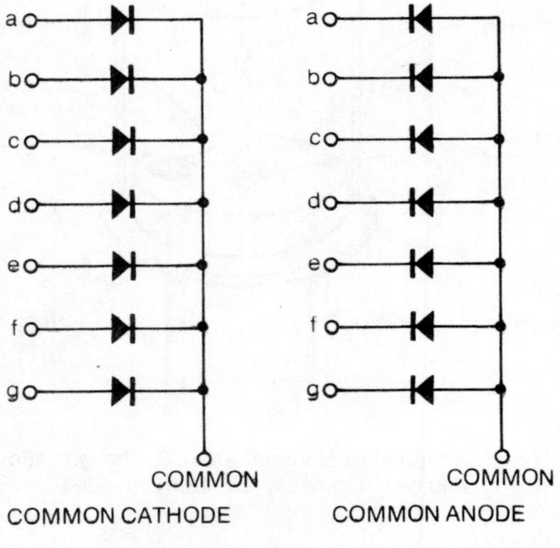

Fig. 3-18. The seven-segment display is made up to an arrangement of LEDs as shown. Each bar can be one, two or three individual LEDs. All the numbers and some letters can be generated. The letter A for instance, would be all segments except d. The seven-segment display is available with decimal points and with the LEDs wired either common anode or common cathode.

The seven-segment readout, for example, is a module which contains seven LEDs within a single package. (Often an eighth, representing a decimal point, is included.) Figure 3-18 shows the arrangement of these seven segments and how they are used to achieve the digits zero through nine. Letters can also be formed with the seven-digit display, but for letters either an eleven-segment display of a five by seven array is used.

In seven- and eleven-segment displays, all of the LED cathodes or anodes are tied together. This permits these displays to be packaged in 12 or 14 leaded, dual in-line packages (DIPs) and, therefore, are compatable with the other packages normally used in integrated circuitry. A five by seven display, though, contains thirty-five LEDs. If they were connected in a similar manner, a one-digit display would require the minimum of a thirty-six pin package. To minimize the number of pins on the package, the LEDs are arranged in a matrix as shown in Fig. 3-19. With this arrangement only twelve leads are required to access all thirty-five LEDs. For example, if a positive voltage was applied to terminal 1 and terminals A through E were grounded, then the entire top row of LEDs would be energized. Similarly, if only E was grounded and 1 through 7 were energized, the entire right-hand column would light. Individual LEDs can be energized by choosing the pair of terminals corresponding to its position in the array.

All the alphanumeric characters can be generated with the five by seven array of LEDs. As an example, consider the letter z. It is easy to see how to generate the top and bottom bars by energizing 1 and 7 and grounding A through E. But if we try to generate the diagonal portion at one time, all the LEDs become energized, since it involves all the lines and columns of the five by seven. To get around this problem, the display is scanned or strobed in either the horizontal or vertical direction. This means that only one row or column at a time is generated. By sequencing through all the rows or columns quickly (100 times per second or faster) it appears to the eye that the entire display is continuously energized in the desired state. Figure 3-20 shows how the letter z is generated by both horizontal and vertical strobing. From one signal, the

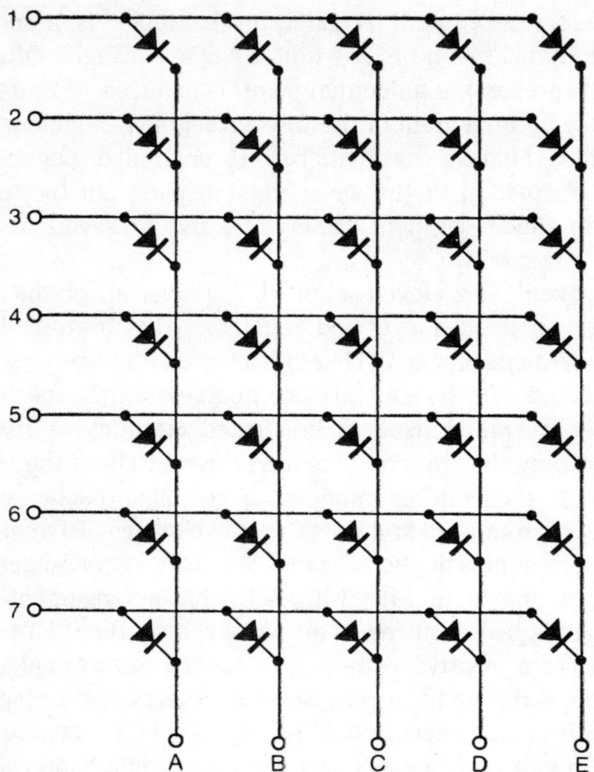

Fig. 3-19. The 5 by 7 array contains 35 LEDs arranged as shown. By applying voltage between different row and column terminals, each individual LED can be energized. Since this array only requires 12 pins, a decimal point can be connected to 2 additional pins and the circuit can still fit in a standard 14-pin DIP package.

letter z has to be changed into the five or seven signals in sequence. This requires additional electronics in the form of ROMs (read-only memories) and a pulse sequencing system.

For the experimenter or hobbyist it is fairly important to get components at a reasonable cost. The lowest cost multiple-digit displays available are those which have been used widely in pocket calculators or watches. Pocket calculator displays have from six to fifteen digits and most have decimal points between each digit. The watch displays are usually four digits with a much smaller character. Calculator characters are usually from 150 mils to 250 mils

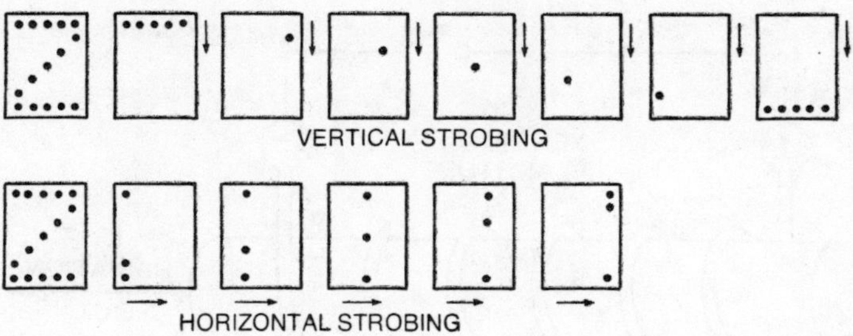

Fig. 3-20. With the 5 by 7 array it is necessary to strobe the display to create the letters. The letter z is shown here for both horizontal and vertical strobing.

high, while the watch characters are usually around 100 mils high. Both display types are usually set for multiplexing. (That is, one set of segment lines, seven segments plus decimal point, and then a digit select line for each digit.) This presents little problem since there are many ICs available for this type of operation. In fact, most calculator and watch chips are designed for this type of operation, as is discussed in Chapter 5. This eliminates the need for any additional electronics in these applications. Most watches and calculators require only a single IC chip.

With these displays the current-drive per segment is usually lower than for discrete LEDs, and usually runs from five milliamps to less than one milliamp for the smaller watch displays where battery life is of prime importance. These lower drain displays are especially useful when interfacing a display with CMOS ICs.

Lasers

Although most of us have heard of lasers, the laser is a recent invention. The first laser was in operation on May 14, 1960, by Dr. Theodore Maimen at Hughes Aircraft Company in Culver City, Calif. Dr. Maiman's laser system is shown in Fig. 3-21. It consisted of a small ruby rod with both ends polished very smooth. The two ends were silvered, like a mirror, with one end having a very thin silver layer so that some light could get through. A xenon flashtube was wound around the rod. Let's see how this laser works.

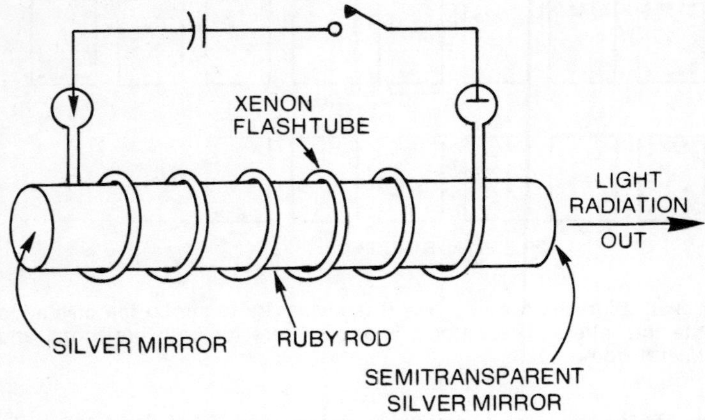

Fig. 3-21. The first laser was a ruby laser similar to the one shown above. Since it was first developed, many other materials have been used in lasers.

When the flashtube is energized, a very large number of photons are generated within a few microseconds. Because the tube is wrapped around the ruby rod, a great percentage of these photons are absorbed by the rod. The absorbed photons excite the ruby, putting many of the atoms in a high energy state. As these atoms give off their excess energy other photons are generated. At the silvered end, the photons which are moving from end to end in the ruby are reflected back and forth. As they move, they strike other atoms and create more photons. More and more photons are generated, increasing the strength of the radiation within the ruby. Since the one end is semitransparent, some of the photons escape at that end in a beam of the same diameter as the ruby. The most important fact about this beam is that the frequency of the photons is confined to the very narrow frequency which is characteristic of the ruby material. This single-frequency light is called monochromatic light.

There are several reasons why the laser was an important development in the field of optophysics. The first is that lenses tend to bend different light frequencies by different amounts. The differences are small, but enough to cause great difficulties when trying to point a light beam great distances or at very small spots. The second is that with a single frequency, the light beam could be used like a radio wave. It would be possible to modulate the beam, as in a radio transmitter, put information on the beam and transmit the information. With multiple frequencies it has been impossible to focus the beam for any distance. Since the beam can be pointed, like microwave transmission, many such light beam transmitters could be used. The super-high frequency of light makes it possible to put many more signals on each beam. It has been estimated that all the telephone calls, radio and television programs, and every other kind of communication taking place in the world today could be transmitted by a single laser. In the very short span of time since Dr. Maiman's first successful operating laser was demonstrated, several hundred laser transmission systems have been put into operation.

The ruby laser, although the first, is not the only type of laser. Many other solids, liquids, and gases can be used. Some work in similar modes while others can be excited directly with electricity, without the xenon lamp. Since each material has its own characteristic frequency, it is possible to purchase or build lasers for most frequencies, from the infrared through the ultraviolet portions of the spectrum

The ruby and other single-crystal types of lasers are excited by strong light. The light sources available for this task are all pulse types. That is, they put out many pulses of high intensity light. For applications such as data transmission, it is necessary to have sources of continuous light beams. This is accomplished in liquid and gas lasers. In these devices an electric current is used to excite the atoms within the optical cavity. This provides a continuous source of excited atoms and therefore a continuous beam of light. Many of these lasers can be modulated directly by the electric current; other lasers are modulated by electronic filters.

There are already thousands of lasers being used in thousands of applications. Medical uses of the laser include welding torn parts of the eye back into place, cauterizing wounds, and burning out cancerous cells. Some other medical uses include the removal of skin blemishes, scars, and tattoos. These applications are all made possible by the ability to control the size and power of the beam. The beam can be focused to a spot on the surface of the skin, giving a very high energy concentration—as high as the surface of the sun—in spots only a few thousandths of an inch in diameter, while only a very small distance away the energy is no stronger than that from a standard light bulb.

Industry has also taken to the laser. Industrial uses include the drilling of super-small holes in materials such as diamond, the hardest substance known to man. Lasers are routinely used in the semiconductor industry to separate the individual devices in a silicon wafer. Before the laser came into use, it was standard practice to allow about 0.010 inches between devices so that they could be separated. In a small diode or transistor wafer, this area could be half of the total silicon area. With laser scribing this distance can be reduced to 0.003 inches, allowing many more devices per wafer. Lasers are also used for welding very small parts together, or welding in hard-to-reach spots by simply bending the beam with a mirror.

Apollo 11 carried to the moon a special mirror reflector. Using a laser and this reflector, the distance to the moon was determined to an accuracy of less than five feet. When the laser beam reached the moon it had spread out to a diameter of only 2.5 miles—quite remarkable in view of the distances involved. The use of the laser for ranging and direction finding will be very necessary for future space activity, as well as being beneficial in air traffic control. Possibly in the future it will be an integral part of an automated highway system where your car drives itself.

The advantages of precise direction and distance measurements are obvious in surveying, mapmaking, and even the construction industry. Lasers were used, as an example, to insure that the tunnel between San Francisco and Oakland for the Bay Area Rapid Transit System was dug in the precise direction desired, eliminating the chance of cutting away excess earth.

A final example which has not reached commercial reality yet, but appears to be just around the corner, is in the computer memory area. Using microminiature solid-state laser arrays, in conjunction with light-sensitive materials, it appears possible to store and retrieve up to a billion bits of data from an area only one centimeter square (about 0.2 square inches). By comparison, the density of most modern computer memories is less than several hundred thousand bits per square centimeter. This is a density increase of nearly ten thousand times.

Chapter 4

Devices Which Use Light

In Chapter 2 many light-sensitive devices were described, and in Chapter 3 light-emitting devices were described. There is a third class of optoelectronic devices which use light as an integral portion of their operation, the prime example being the optocoupler. A fourth class of optoelectronic device is that which by the use of electrical fields can vary their optical properties. Of this class most readers will be more familiar with the liquid crystal displays.

OPTOCOUPLERS

Photocoupler, photon-coupled isolator, optoisolator, and optocoupled isolator are all names for the same family of components. These components are analogous to transformers in that they are used to transfer signals from one part of an electrical system to another, while maintaining electrical isolation. The transformer uses a magnetic field which is contained within the ferrous core to transfer the energy from one system to another. The optocoupler (that's what it will be called here) uses light energy to transfer the information. A standard transformer will transfer over 95% of the energy from one circuit to another while the optocouplers typically transfer less than one percent.

Even though the energy transfer is very low for the optocoupler, it offers distinct advantages in many applications over transformers. The first and probably the most important advantage is its ability to transfer DC signals as well as AC signals. Second is its size advantage. Many couplers are packaged in TO-5's or six-pin DIPs. Third is the compatability of many optocouplers with digital integrated circuits, and in fact with most solid-state electronics. The fourth factor is the low cost, both of the optocoupler and optoelectronic system, over the transformer and the electromagnetic system. The place where optocouplers have the advantages over transformers is where information rather than power is transferred, and where low frequency information, even DC, is transmitted.

The optocoupler contains three main components, as shown in Fig. 4-1. A light source is connected to the input terminals, and a light-sensitive device to the output. These two components are physically separated by one or more electrical insulators. The insulators, though, are optically transparent to some wavelengths within the band of frequencies common to the light emitters and the light-sensitive device's spectrum.

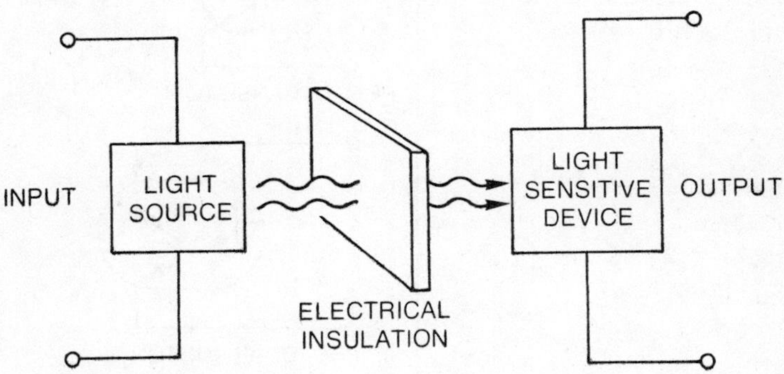

Fig. 4-1. The basic optocoupler consists of a source of light, an electrically non-conductive but optically transparent light path, and a light-sensitive device.

Of the light emitters available, three types find use in most optocouplers. Each of these has different characteristics suited for certain applications. These are shown in Fig. 4-2.

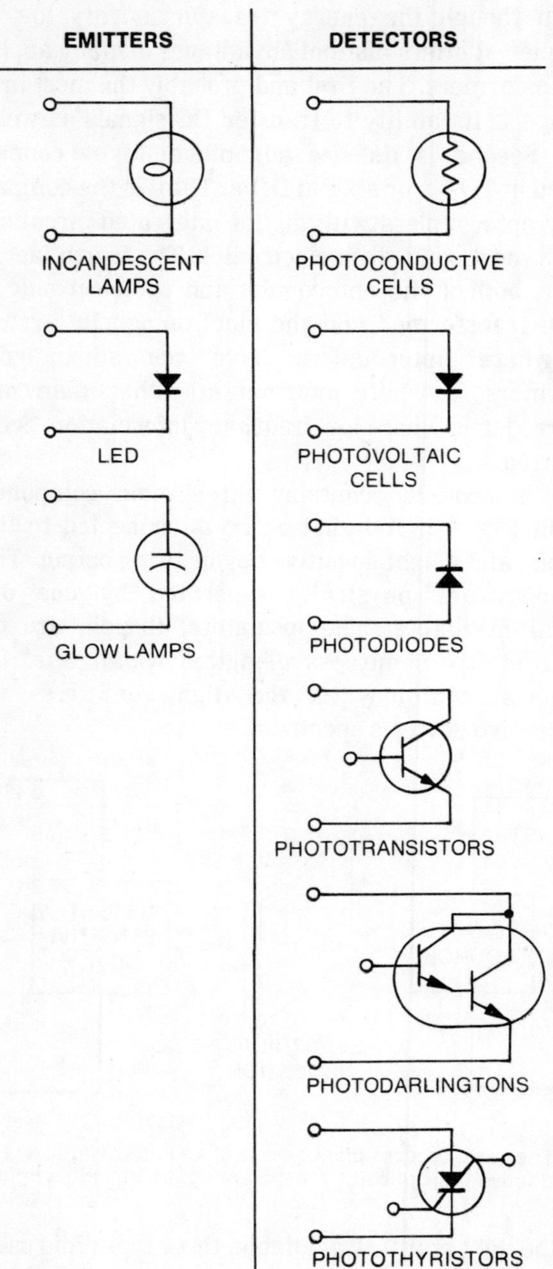

Fig. 4-2. The common emitters and detectors used in optocouplers.

The incandescent lamp has a much slower response time than either the LED or glow lamp. This makes it suitable for average voltage applications. The LED is suitable for providing instantaneous values, and the glow lamp is primarily used as an on/off indicator.

The different light-sensitive devices used in optocouplers are also shown in Fig. 4-2, and their characteristics described in Table 4-1. All the devices shown are semiconductor devices. The photoconductive cells are made from several different semiconductor materials, so it is possible to tailor their spectral response to the light emitter that will be used. The balance of the light detectors are basically silicon devices and are sensitive mainly to the near-infrared. This is the reason glow lamps are usually used only with photoconductive cells. The devices shown in Fig. 4-2 are explained in Chapter 2. The different light emitting and detecting devices combined into an optocoupler determine the characteristics of the device.

LIQUID CRYSTAL DISPLAYS

Seven-segment displays using LEDs were discussed. An LED display such as this requires about ten milliwatts per segment. This creates problems with battery life in miniature battery-operated devices. In a watch for example, there are 23 segments required for a normal 12-hour display of hours and minutes. When the display is energized it dissipates 230 milliwatts. Though this is small, the typical watch battery only has about 250 milliwatt-hours of capacity. It is easily seen that the display will use much of the battery capacity. If the watch manufacturer could use a display that only required microwatts, the battery life could be extended considerably. The liquid crystal display meets this requirement.

Figure 4-3 shows how a typical liquid crystal display is assembled. The display consists of a two-piece transparent case with the liquid crystal material filling the space between. Very thin electrodes are deposited on a transparent material (such as tin oxide) on the inner surface of both the base and the cover. The electrodes are shaped to give the desired segment shapes. The exterior of either the cover or the base may be coated with a mirror-like material. In other displays, a

Table 4-1. A Comparison of Some Common Optocouplers. The Speed and Sensitivity Are Ranked Numerically With One Being the Highest.

Source	Incandescent Lamp		LED				Glow Lamp
Detector	Photo-conductive cell	Photo-voltaic cell	Photo-diode	Photo-transistor	Photo-darlington	Photo-SCR	Photo-conductive cell
Speed	7	6	1	2	4	3	5
Sensitivity	5	4	6	2	1	3	7
Polarity Sensitive Input	no	no	yes	yes	yes	yes	no
Polarity Sensitive Output	no	yes	yes	yes	yes	yes	no
Input Voltages	3–120	3–120	1.5	1.5	1.5	2.5	80–150
Output Voltages	≤250	0.5–1.0	≤25	≤25	≤25	≤400	≤250

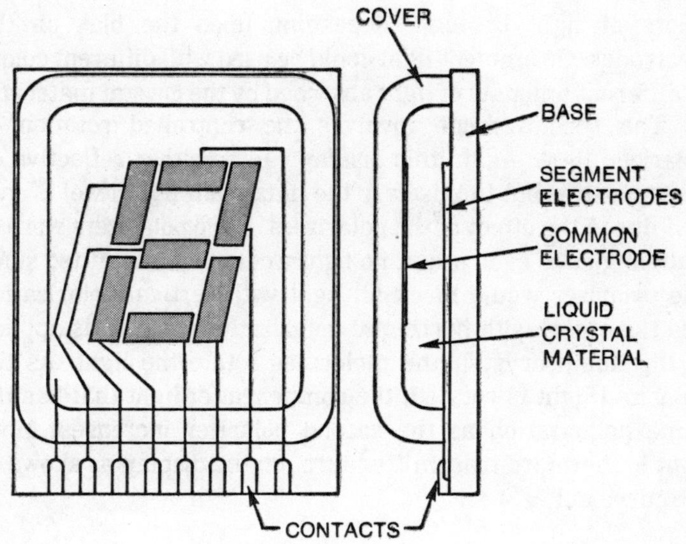

Fig. 4-3. A basic liquid crystal display. When an electrical field is applied between the common electrode and any of the segments, the liquid crystal material between changes its optical properties.

polarizing film is put on both of the exterior surfaces with the polarization at defined relationships to each other.

Liquid crystals are liquids in which the intermolecular bonds have definite relationships, like in crystals. The field of liquid crystals is just beginning, but there are already many materials which can be used in displays. The different materials will give different effects and several types of displays.

Basically, when a liquid crystal material is subjected to an electric field, its optical characteristics change. In some materials the field is AC and in others DC. Four main phenomena can occur in liquid crystals when subjected to an electric field.

The first mode of operation of a liquid crystal is a change in the optical density or color of the material. When the optical density increases, less light can pass through the liquid crystal display. Displays of this type can be either reflective or transmissive. This means a lamp could be placed behind the display and the display would allow different amounts or colors of light through, depending upon the bias on the electrodes. Or ambient light could be used with different colors or different amounts of light absorbed by the crystal material.

The second mode involves the controlled rotation of polarized light. Again this display can be either reflective or transmissive, but in this case the maximum light level is only 50% due to the effect of the polarizers. If the polarizers were at right angles to each other, no light would be transmitted since one polarizer would block all light with vertical polarization and the other, with horizontal polarization. As bias is applied to the liquid crystal, the molecules rotate the light. As the polarized light is rotated, the component of light that has the same polarization as the second polarizer increases. More light is therefore transmitted through the display as shown in the curve in Fig. 4-4.

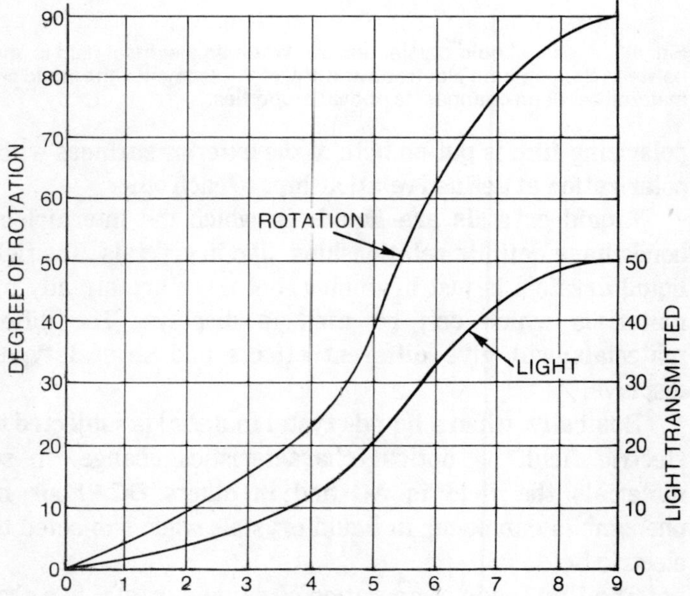

Fig. 4-4. With a liquid crystal of the controlled rotation type, the amount of light transmitted changes with the bias voltage. This type of unit can therefore be used as an electronic filter or shutter.

The third mode for liquid crystals is the change of the index of refraction of the material with an applied electrical field. In Chapter 1, Snell's law of refraction was shown for light rays going from one medium to another as:

$$N_A \sin(\Phi_3) = N_B \sin(\Phi_1)$$

where N_A and N_B were the index of refractions of the two materials, and Φ_1 and Φ_3 were the angles with respect to the perpendicular to the surface. With two materials, if the index of refraction of one is significantly different from the other, there are angles where this equation cannot be satisfied. (Those where the sine of Φ_3 must be greater than one to satisfy the equation.) For these angles the light will be reflected from the surface rather than transmitted. (Looking into a pool of water is a good example of this principle. If we look at an angle, we cannot see past the surface, but if we look straight down we can.) Changing the index of refraction would also change the critical angle, as shown pictorially in Fig. 4-5.

Dynamic scattering is the fourth operating mode. With this method, the light entering the liquid crystal is scattered in all directions instead of being transmitted straight through. When a bias is applied to the electrodes, the scattering effect is eliminated and the light is transmitted directly through.

With the different materials and operation modes, liquid crystal displays are finding wide use in electronics and in other uses, such as signs. The size of the liquid crystal display is limited only by how large of a case that can be processed for electrodes and sealed. Multiple effects can be achieved by using different materials together, different voltages, or different electrodes. It should be remembered that most liquid crystal displays are slow in their response as compared to LEDs or even incandescent lamps.

FIBER OPTICS

Light can travel great distances, but unfortunately only in straight lines. In many places it is convenient to get light to bend around corners or travel a flexible or arbitrary path. For these applications fiber optics are used. The name describes basically what they are; small fibers of optically transparent material. The fibers available today are usually glass or plastic. Around the fibers, which vary from a few thousandths

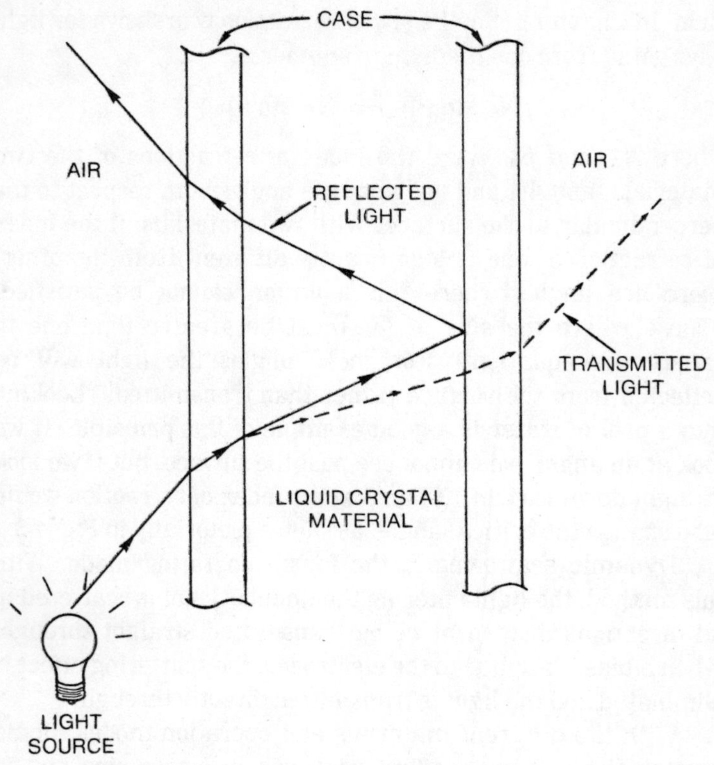

Fig. 4-5. If the index of refraction of the liquid crystal is high, the light beam is bent closer to the surface perpendicular and can be transmitted as shown by the dotted line. When the index of refraction is lower, the light beam is reflected back into the liquid crystal.

of an inch in diameter up, are coatings with a very high index of refraction or a metalized coat. This forms an optical tunnel, where the light in the fiber hits the edge of the fiber and is bounced off the edge back into the fiber, as shown in Fig. 4-6. The plastic fibers are available in cut lengths either individually or in bundles for very low cost at most electronic chains that specialize in serving the hobbyist.

With fiber optics, very high isolation voltage optocouplers can be made with the emitters and detector separated by distances of up to hundreds of feet. It must be remembered though, that like all materials, fiber optics have their own unique spectral response. Some types attenuate infrared which is the important wavelength to silicon and gallium arsenide.

There are finite losses in fiber optics which can be divided into two parts—the end loss and the loss with length. The end loss can be considerable and should not be ignored. To minimize this loss, the end into which the light is introduced should be as flat as possible, and the light should be a parallel beam perpendicular to the end surface to minimize the reflection. Special glues are available to permit securing the fibers to the light emitters to minimize attenuation and reflection.

The light-emitting end can be tailored to the application. If the fiber is to give a point of light, it can be heated with a match to make a small ball on the end. This ball will give a small point of light which is viewable over a wide angle. To use the fibers to shine light on a small light-sensitive device, the end should be made flat as mentioned.

Much development work is being carried out today on fiber-optic systems, primarily in the areas of length and interface losses. Fiber-optic cables with losses of less than two decibels per kilometer have been developed. This makes possible the transmission of light over distances equivalent to most microwave transmission links. These microwave links are used both for telephone long-distance communications and for power-system information and data systems. As was discussed in the laser section, many more phone conversations can be put on a laser beam than on a microwave signal. The fiber-optic system is also not limited to line of sight as is the

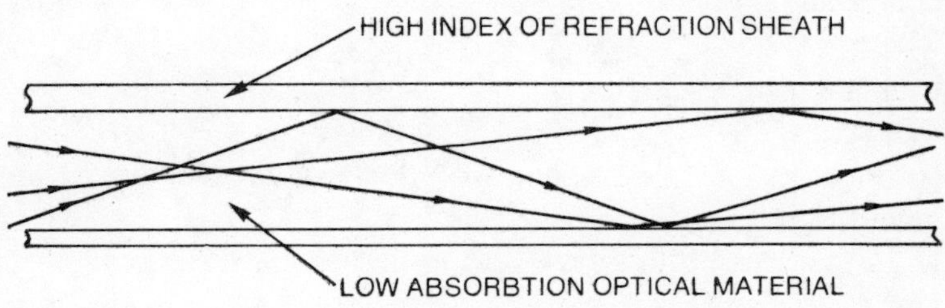

Fig. 4-6. When light travels down the fiber-optic filament, the light is bounced back into the fiber each time it hits the edge of the fiber. This is due to the differences in the index of refraction between the sheath and the fiber.

microwave system. A still unresolved problem is joining one fiber-optic cable to another. The search continues for a connector system which will allow quick and easy joining of two or more fibers end to end. Presently the fibers are ground smooth then carefully glued together using special glues, which is impractical for anything except laboratory work. Several connector systems have been developed, but none have reached a production stage.

Chapter 5

Applications of Optoelectronic Devices

There is now a very long list of applications for optoelectronic devices. It would take several volumes to document in any detail even the most important ones. In some applications, the light is the primary element of the system. Examples of this are lighting control, photographic systems, and automatic nighttime switches. Other applications use the light as the medium by which a function is controlled. In this group are position and rate sensors. In a third group of applications, the optoelectronics are incidental to the system and used only for expedience. These applications encompass most of the optocoupler applications. The following few pages are intended to demonstrate some of what has been done.

LIGHT REGULATION

Figure 5-1 shows one way to maintain a constant light level for an incandescent lamp. The system uses a unijunction transistor Q1, in ramp and pedestal configuration to trigger a triac. The pedestal is controlled by the resistance divider made up of the photocell and the set potentiometer, R5. With light in excess of the set level, the photocell's resistance drops, causing the circuit to trigger later in the sine wave. This, in turn, reduces the power to the lamp. Detailed descriptions of

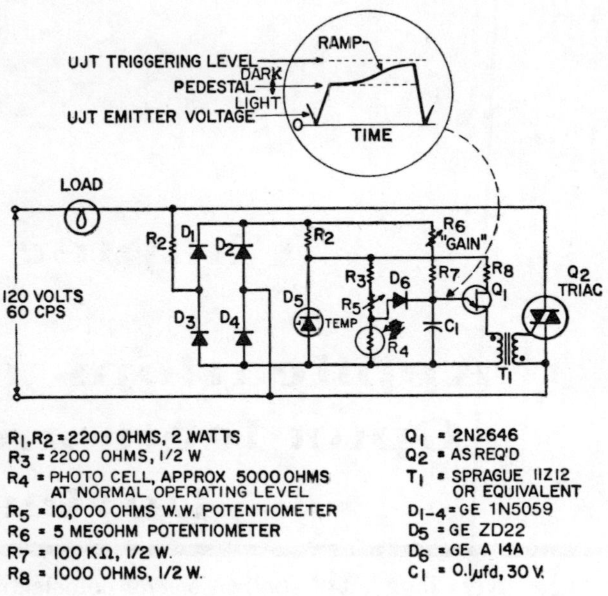

Fig. 5-1. A precision proportional lighting control. If R4 is placed so that it receives light from the load, the circuit will regulate the light level as the in voltage changes. (Courtesy of General Electric, Semiconductor Products Dept.)

ramp and pedestal circuits are contained in TAB book number 695, *Practical Triac/SCR Projects for the Experimenter*.

LIGHT SWITCHES

It is often necessary to energize a load if a light source disappears. The circuit shown in Fig. 5-2 accomplishes this by using a light-activated silicon controlled switch (LASCS) such as the General Electric L1V. When the LASCS is illuminated, the capacitor-charging current is shunted through the LASCS and therefore the SCR cannot be triggered. When the illumination is removed, the capacitor charges on the next positive half-cycle. At about eight volts across the capacitor, the silicon unilateral switch (SUS) turns on and triggers the SCR. The SCR is retriggered on each subsequent cycle until the LASCS is again illuminated. With the components shown, loads up to 18 amperes can be controlled if the SCR is

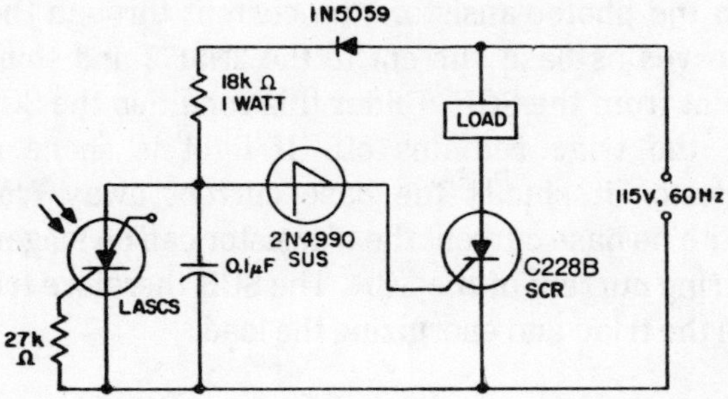

Fig. 5-2. Half-wave light interruption detector. The circuit of Fig. 5-3 performs the similar function on a full-wave basis. (Courtesy of General Electric, Semiconductor Products Dept.)

adequately heat sinked. The voltage applied to the load is half-wave rectified so only DC loads can be used.

The circuit shown in Fig. 5-3 operates in a manner similar to Fig. 5-2 except that it is full-wave control. A triac has replaced the SCR and a diac the SUS. To achieve the bidirectional characteristic in the LASCS, it has been placed on the DC side of a rectifier bridge.

In case you want to turn on a system with light, the circuit shown in Fig. 5-4 will most likely fill the bill. The triac which energizes the load is triggered by the SCR. The SCR is in turn triggered by the current through the 47K resistor. When no

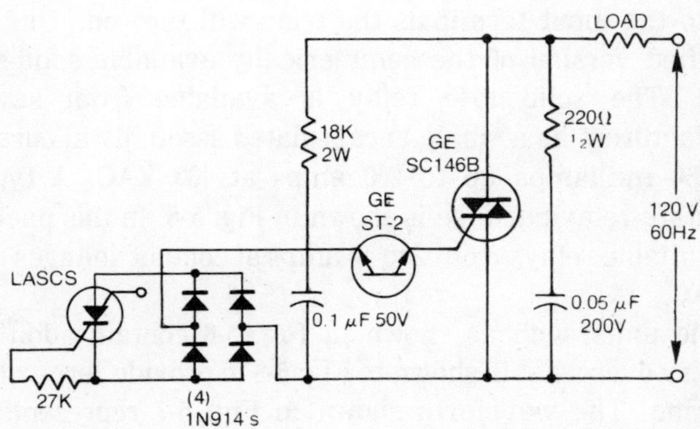

Fig. 5-3. When light is removed from the LASCS, the triac is energized in this light interruption detector.

light is on the phototransistor, the current through the 220K resistor serves as base current to the 2N5172 and shunts the gate current from the SCR. Under this condition the SCR and, therefore, the triac remains off. If light is shone on the phototransistor it shunts the base current away from the 2N5127. With no base current the transistor can no longer shunt the triggering current of the SCR. The SCR therefore triggers, turning on the triac and energizing the load.

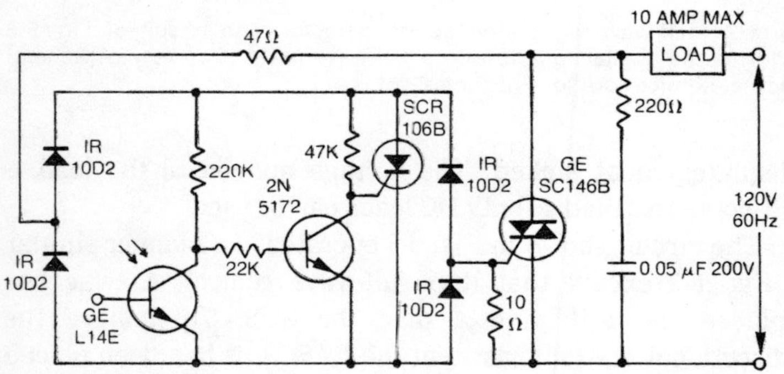

Fig. 5-4. Light on the phototransistor will cause the load to be energized. A variation of this circuit is found in Project 9.

SOLID-STATE RELAYS

By modifying the light-sensor circuitry, as shown in Fig. 5-5, a solid-state relay circuit results. Upon application of 3 VDC to the input terminals the triac will turn on. This is a simplified version of the commerically available solid-state relays. The solid-state relay is available from several manufacturers as a single encapsulated assembly in currents from 50 milliamps up to 200 amps at 480 VAC. A typical solid-state relay package is shown in Fig. 5-6. In this package are available relays from 2 to 40 amps at contact voltages up to 480 VAC.

The units such as shown in Fig. 5-6 contain additional circuitry over what is shown in Fig. 5-5 to provide zero-voltage switching. The waveform shown in Fig. 5-7 represents the zero-voltage switching of a solid-state relay. After the input signal is applied to the relay, the load is not immediately

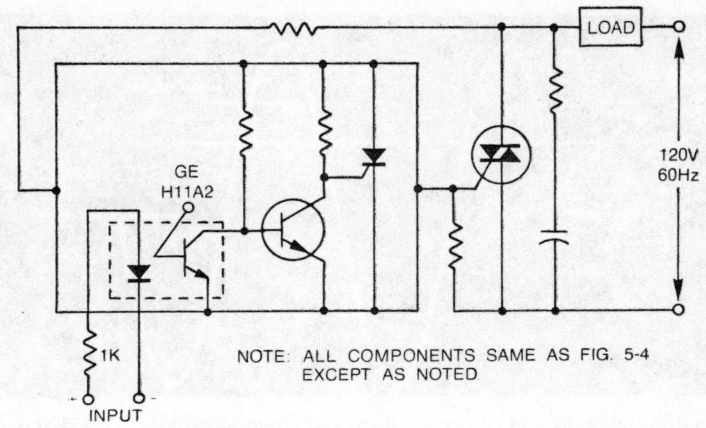

Fig. 5-5. A simple solid-state relay with optocoupling for input output isolation. This circuit is similar to several commercial solid-state relays.

energized as would occur with the circuit of Fig. 5-5, but is energized at the next zero-voltage crossing of the AC line voltage. This action serves to minimize or, even in some cases, eliminate radio-frequency interference. Likewise, when the input signal is removed, the relay remains in conduction until the next load current zero crossing. In this way there are no

Fig. 5-6. The Crydom solid-state relays provide the ability to control up to 40 amperes at 240 volts AC from a TTL input circuit. These units measure only 1.75 by 2.25 by 0.87 inches. (Couresty of Crydom Div. of International Rectifier.)

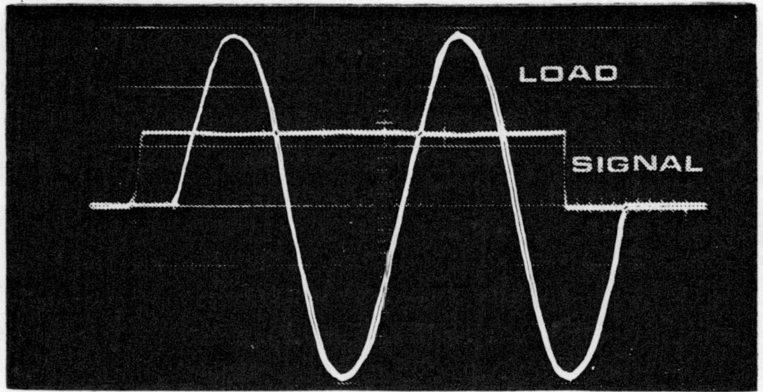

Fig. 5-7. The addition of zero-voltage switching eliminates the sharp discontinuity in load voltage waveforms. This is one of the main advantages of solid-state relays. (Courtesy of Crydom Div. of International Rectifier.)

transients or arcing that normally occur when switching an electromechanical relay.

The lack of transients, the long life, and the many switching operations of solid-state contacts make the solid-state relay ideal for applications requiring a high rate of repetitive switching, or where the inductive load ratings of an electromechanical relay prohibit their use. Several of these applications are discussed in the following paragraphs.

Although solid-state relays are not self-latching relays, a simple circuit as shown in Fig. 5-8 allows the solid-state relay

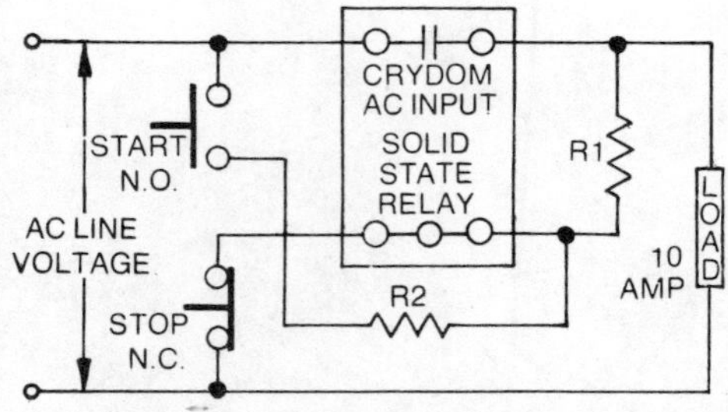

AC LINE VOLT.	R1	R2	SS. RELAY
120	4.7K	0	A1210
240	47K	0	A2410
480	47K 2W	47K 2W	A4812

Fig. 5-8. The solid-state relays can be used as pushbutton latching relays as shown. (Courtesy of Crydom Div. of International Rectifier.)

to latch on. The second pushbutton serves to turn the relay off. The pushbutton switches can be replaced with triacs or SCRs if electronic control is desired, although other solid-state relay circuits can accomplish the electronic switching easier.

A major application of solid-state relays is motor control. The circuit of Fig. 5-9 shows solid-state relay reversing drive. The motor used is a balanced-winding split-capacitor motor similar to those used in some electric garage door systems. By energizing the inputs of one or the other solid-state relay, the motor will run either forward or reverse. The resistor is included to limit the capacitor discharge in case both relays get accidentally energized at the same time. Since the *off* relay is across an LC circuit, it can see voltages up to twice the supply voltage. This must be considered when choosing the solid-state relay.

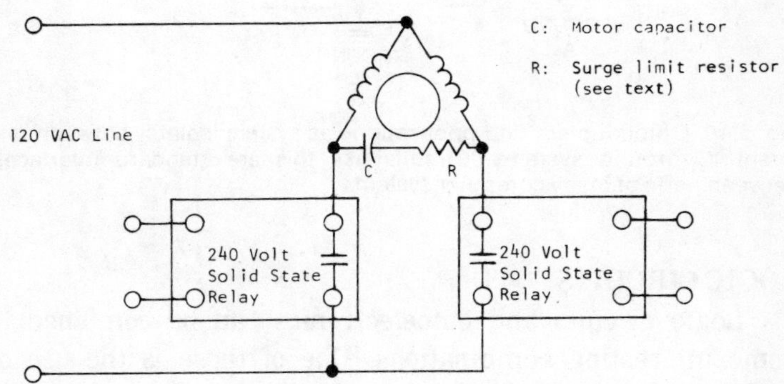

Fig. 5-9. Solid-state relays find many applications for controlling motors. This reversing drive is a typical example. (Courtesy of Crydom Div. of International Rectifier.)

Solid-state relays have found applications in many types of equipment—from traffic controllers to swimming pool controls to office and computer equipment.

In computer equipment, solid-state relays are used to turn on and off drive motors for tape transports and disc drives. Other optoelectronic devices are also used in these systems. LEDs are used as status indicators on the operator consoles. Photoconductive cells or phototransistors are used in punched-tape and card readers. One of the most interesting applications for optoelectronics in computers is optocouplers as interface elements between different machines in a computer system. Since there may be differences in ground potentials between two adjacent machines, it is necessary to electrically isolate the machines but to allow the data to be transferred. By using optocouplers as shown in Fig. 5-10 this is easily and simply accomplished.

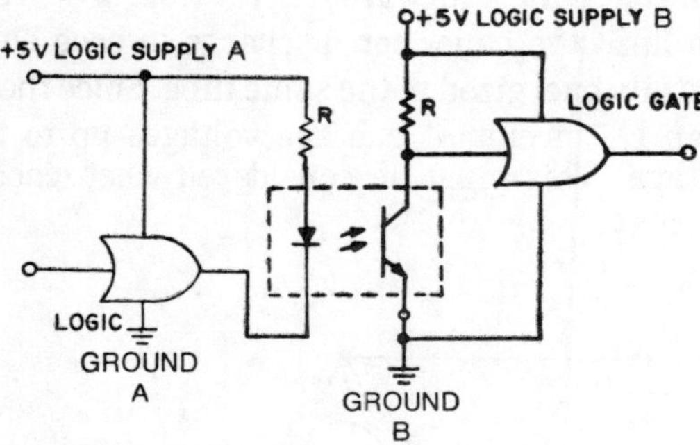

Fig. 5-10. Optocouplers find applications as system isolator between different IC ground systems. Circuits like this are standard interfaces between parts of many computer systems.

LOGIC CIRCUITS

Logic circuits and optoelectronics can be combined in some interesting combinations. One of these is the use of multiple optoelectronic devices to perform the sensing and logic at the same time. Five combinations like this are shown in Fig. 5-11. Circuits (a) and (b) show *AND* circuits, while (c) and (d) show *OR* circuits. Circuit (e) gives a light-triggered flip-flop. Each of these circuits can be expanded to more inputs. Flip-flops can be expanded to either a chaser circuit or a ring-counter circuit, depending upon how the capacitors are connected between the SCR anodes. The circuits using SCRs shown in Fig. 5-11 (a) and (c) can be made to latch continuously using a DC supply voltage or momentarily by using an AC supply voltage. The transistor circuits, of course, should only be used with a DC supply voltage.

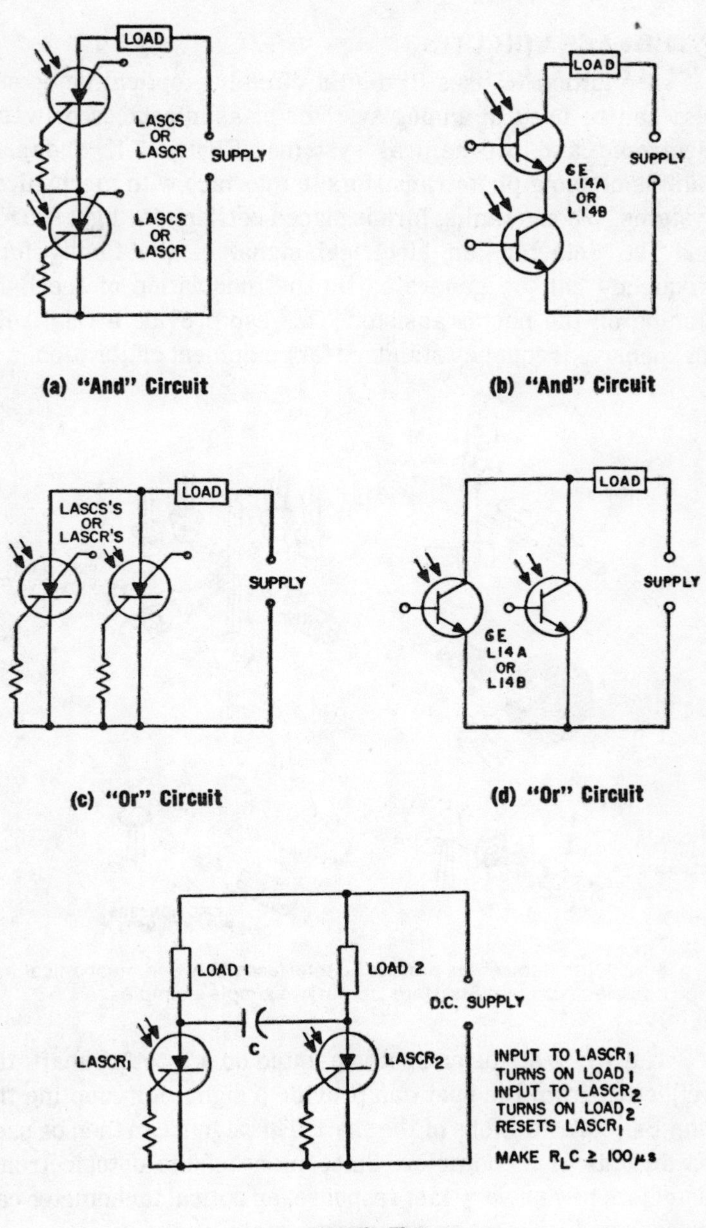

Fig. 5-11. Five simple input logic circuits. The thyristor circuits of (a) and (c) will latch on if used on a DC supply voltage. (Courtesy of General Electric Semiconductor Products Dept.)

INTERFACE CIRCUITS

In addition to uses in digital circuitry, optical detectors also can be used in analog systems or as interfaces between electronic and mechanical systems. Figure 5-12 shows a sample of using phototransistors to interface with mechanical systems. When a tuning fork is placed between the light source and the detector, an electrical signal of the tuning fork frequency can be generated by the modulation of the light shining on the phototransistor. This can provide a relatively inexpensive frequency standard for equipment calibration.

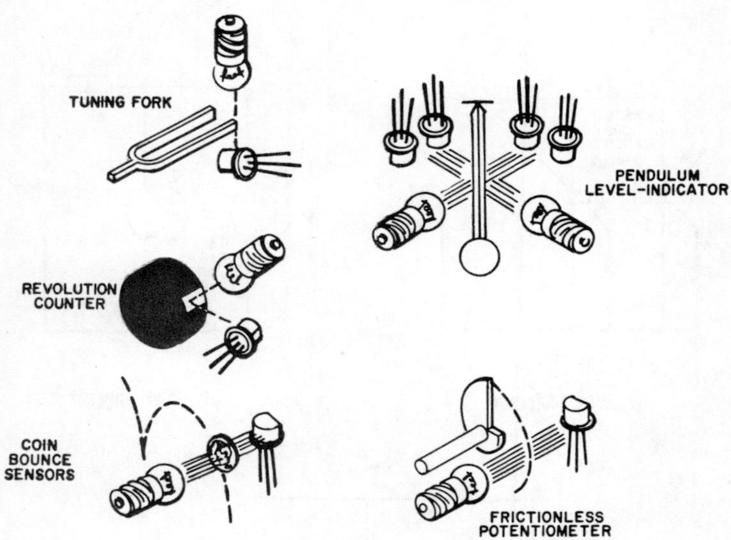

Fig. 5-12. Optoelectronics are useful interfaces between mechanical motion and electrical circuits. Here are but five simple examples.

If a white stripe or spot is painted on a rotating shaft, the reflection from the spot can provide a signal for counting the number of revolutions of the shaft. This signal can then be used as an optical tachometer. Since many of the optoelectronic detectors have a very fast response, an optical tachometer can be designed for most any rotating system.

The coin-bounce sensor is a specialized example of units which detect the presence of an object. With the phototransistor connected to the coin counter, the amount of money deposited can be determined. You will note that a washer will not trip the properly designed detector system because of the light through the hole in the center of the washer.

If you wish to design an electronic level, the pendulum level-indicator can be the basis on which to start. If the pendulum is off center, one of the phototransistors will receive more light than its mate. This signal can then be fed into a bridge circuit or a differential-amplifier circuit which can indicate the direction of imbalance.

The frictionless potentiometer solves the problems of noisy potentiometers. Instead of a wiper, a variable shutter reduces the light transmitted from the lamp to the detector. A helical-shaped shutter has been found to work very well.

Transparent plastic discs with printed opaque patterns have been used in several applications. Figure 5-13 shows a pattern which can be used to generate sound. When a light source is placed on one side of the disc and a detector on the other, the detector will receive more or less light depending upon the width of the transparent section. When the disc is rotated at a reasonable speed, the modulation of the light received when amplified will give the desired frequency. The

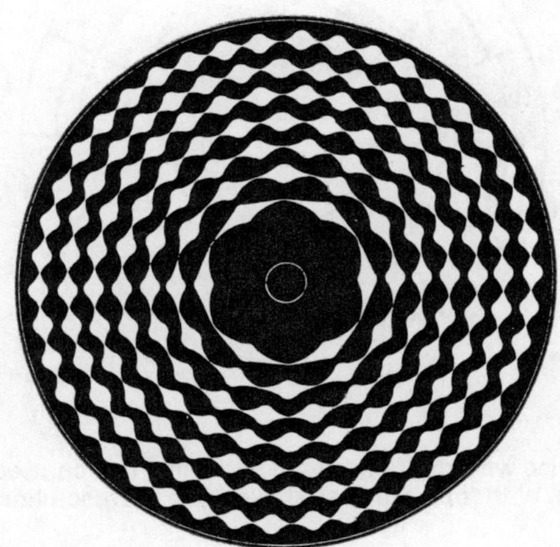

Fig. 5-13. With light sources and detectors on both sides of a transparent disc with printed patterns such as this, sounds can be generated. The text explains some of the uses for such systems. (Courtesy of International Rectifier.)

other bands represent harmonics of the base frequency. By providing each band with its own emitter/detector pair, the relative intensity of each harmonic can be controlled independently. In fact, any waveform can be produced by photographing its oscillograph picture and then reproducing it in a circular format around the disc. This is the manner in which several of the less expensive rhythm attachments for electronic organs operate.

Rotating discs can also be used to display the position of the shaft by using a binary-coded disc. Depending upon the accuracy desired, the disc is divided into sections. In Fig. 5-14, a disc has been divided into 16 sections. Each section has been binary coded by the four white bands. (Check segment 15 to see all four bands.) By feeding the resulting count into digital circuitry the position of the shaft or disc is determined. With only 4 bands, 16 segments are used; with 5, 32 segments, and so on. Nine bands are required to be able to position the shaft and disc to less than one degree of rotation.

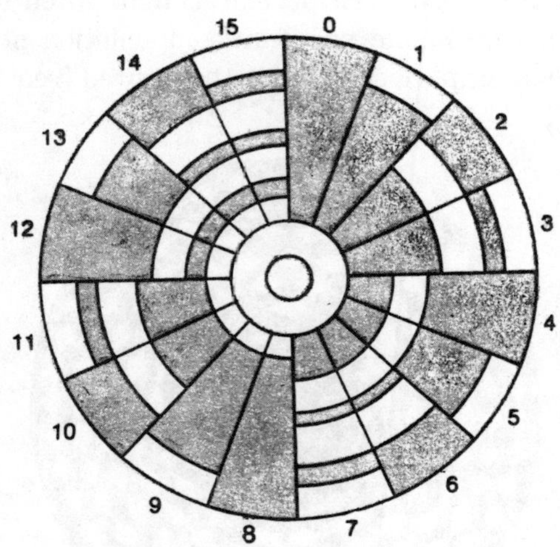

Fig. 5-14. A disc with digitally coded segments can be used to determine shaft position. With four bands as shown here the resolution is only 22.5°.

In industrial applications it is often necessary to trigger series strings of high-power SCRs. Trigger transformers can not be used in these applications because of the high voltages involved. In recent years, systems such as shown in Fig. 5-15 have gained in popularity. In this system, a xenon flashtube provides a triggering signal to each of five light-activated SCRs via a fiber-optic path. The flashtube turns off quite quickly and provides an extremely high-energy burst of light. This light then travels to the triggering LASCR via the fiber optics. When the LASCR triggers, it, in turn, triggers the high-power SCRs. This system has the advantage of completely isolating the control circuit (associated with the

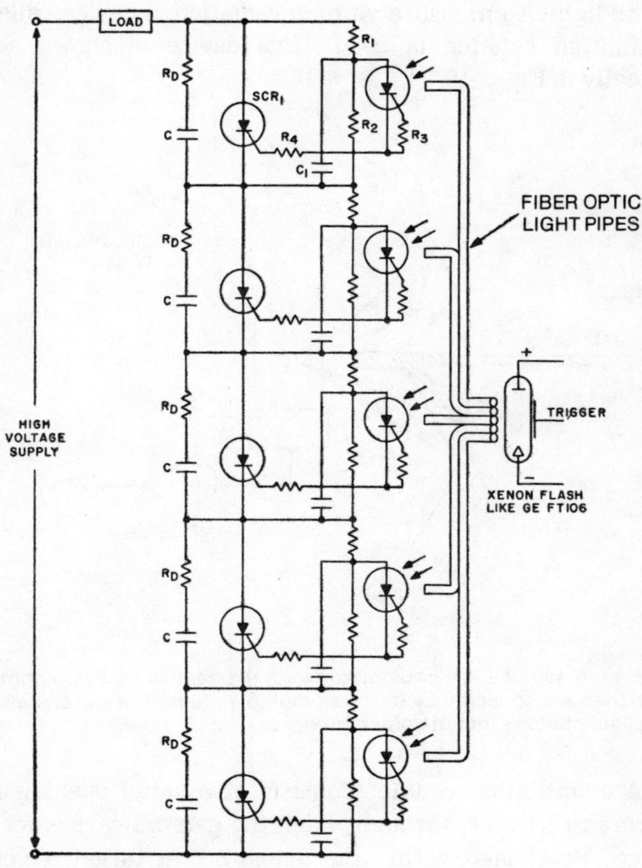

Fig. 5-15. SCRs in high-voltage applications can be triggered using a combination of light-activated SCRs, fiber optics, and a xenon flashtube as shown here. (Courtesy of General Electric Semiconductor Products Dept.)

flashtube) and the power system. There is absolutely no way to get transients and electrical noise from one subsystem to the other.

On the other side of the world of science, there is need to measure radiation of super-short wavelengths such as gamma rays and X-rays. This is accomplished by converting the radiation into longer wavelengths in a manner similar to that of a fluorescent lamp. In the fluorescent lamp, the ultraviolet radiation from the mercury vapor is converted into visual light by the phosphor coating on the inside of the tube; we can then see the light. To measure gamma radiation, a device called a scintillation counter is used. This device is shown schematically in Fig. 5-16.

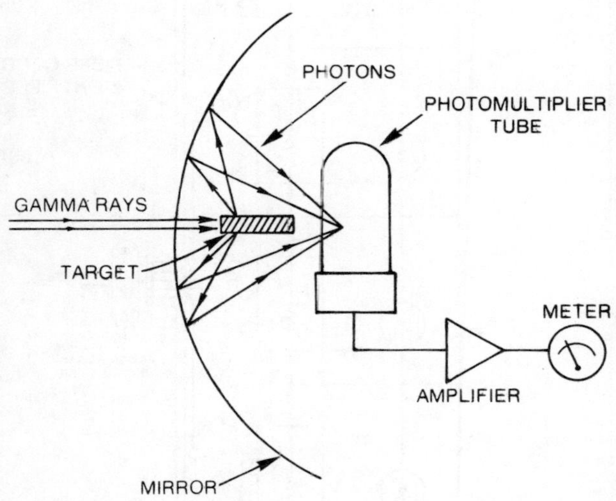

Fig. 5-16. A scintillation counter converts the gamma rays into photons, which then are collected by the photomultiplier tube. The mirror serves to reflect the photons into the photomultiplier tube.

A scintillation counter consists of a small opening in a mirrored reflector, through which the gamma rays enter the device. Positioned within the opening is a target which is usually made of cesium iodide with a slight doping of thallium or another of the inorganic phosphors. When the gamma rays strike the phosphor, they are absorbed by the atoms, elevating the electron energy levels. As the electrons return to their lower energy levels, they emit photons. These photons are then collected by the photomultiplier tube. By proper positioning of the target, mirror, and the tube, almost all of the photons can be collected. The output of the photomultiplier tube is then amplified, and the resulting signal is used to drive a meter or recording device.

By the simple changing of targets, an instrument can be made to measure high-energy radiation of most any type.

Some electronic systems require the collection of radiation to power the system. A popular means of accomplishing this is by the use of solar cells. This can be done quite easily by series and parallel strings of these devices (see Fig. 5-17). A single silicon cell has an output voltage of about 0.4 volt in sunlight, while the selenium cell has about 0.25 volt output. The desired output voltage determines the number of series solar cells required. If the load is known, the power generation needed will give the total area of solar cells needed. A rule of thumb is to use 0.1 watt per square inch for silicon cells and 0.05 watt per square inch for selenium. The number of parallel strings can be determined from the area required. If it is necessary for the system to operate over a twenty-four

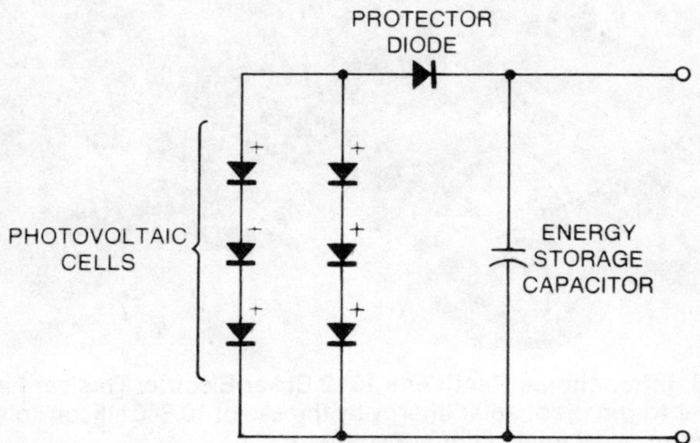

Fig. 5-17. Solar cells convert light into electricity. Placing the individual cells in series gives the required voltage, and paralleling the strings gives the required output power. The protection diode prevents reactive loads from damaging the cells. The capacitor serves to even out the load presented to the cells.

hour period, the storage capacitor must be replaced by a rechargeable battery pack. It is also necessary to multiply the size of the solar-cell array by the hours of daylight and factors for the sun's angle to the solar-cell array to insure adequate output.

An interesting example of solar cell application is shown in Fig. 5-18. Several years ago, International Rectifier converted the 1912 Baker Electric car shown to run off of solar energy. A 4×6.5-foot array of silicon solar cells was added to the top of the car. The array consisted of 10,640 1×2-cm cells. The array generated approximately 100 watts at 115 VDC open circuit. The array weight is only 45 pounds, not as much as two automobile batteries.

Fig. 5-18. International Rectifier's 1912 Baker Electric. This car has been converted to run off of solar energy by the use of 10,640 silicon solar cells. (Photo courtesy of International Rectifier.)

Light-emitting diodes find many uses as indicators. They are used on almost every minicomputer as status indicators. In addition, they are finding large usage in other types of systems. Figure 5-19 shows a simple circuit which will monitor a five-volt power supply for TTL ICs. The LED will be on only for power-supply voltages within the acceptable TTL limits. This circuit can come in very handy in laboratory or shop work on TTL circuits. The system consists of a pair of transistor level detectors. The first (Q2, Z1, R2, R2) turns on the LED if the voltage applied exceeds 4.5 volts. If the voltage is increased to 5.5 volts, the second transistor (Q2) is turned on by its voltage-sensing network (Z2, R4, R5). The turning on of Q2 removes the base drive from Q1. This turns off Q1 and the LED. The system can be modified for almost any voltage level through the proper choice of zeners.

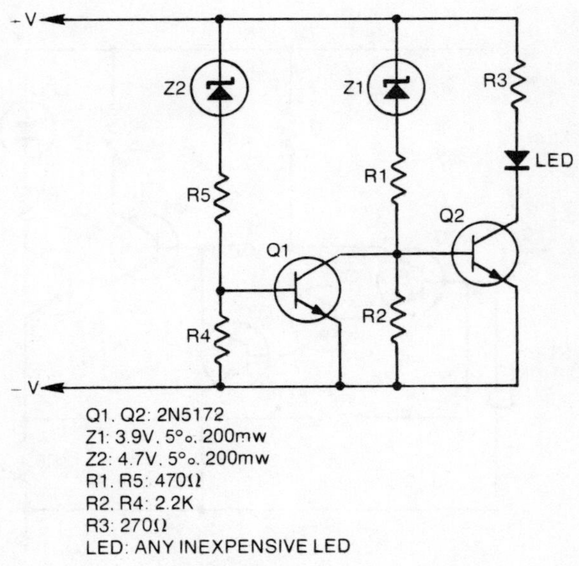

Fig. 5-19. An LED voltage level indicator. The LED is on only for supply voltages between 4.5 volts and 5.5 volts.

LIGHT-TO-SOUND CIRCUITS

Project 2 shows a lamp bulb being modulated by an audio signal to make a light-beam transmitter. The lamp's frequency response is limited to a couple of kilohertz by the mass of the filament which must heat and cool in response to the sound input. An LED is not limited by thermal response times, and, in fact, some GaAs LEDs can be modulated at frequencies of up to 100 MHz. The circuit shown in Fig. 5-20 is designed to modulate the LED at audio frequencies, while the circuit shown in Fig. 5-21 can modulate the LED's output at radio frequencies. Either circuit can be used with GaAs laser diodes in place of the LEDs to increase the range of the beam.

The circuit of Fig. 5-20 consists of basically a two-stage amplifier. Q1, Q2, and Q3 are connected in a triple darlington configuration and provide a very-high-gain amplifier stage. Q4 and Q5 provide the second amplifier stage. R4 and R5 provide the current limiting for the output LED, while R6 is a feedback resistor which sets the quiescent current level as well as the total loop gain.

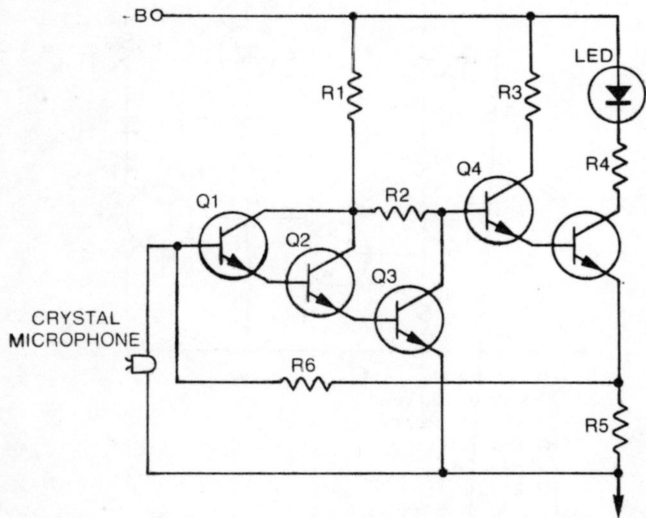

Fig. 5-20. An LED's output can be modulated with audio frequencies, using the crystal microphone amplifier circuit shown here. This circuit can be used in place of the light beam transmitter of Project 2.

The circuit in Fig. 5-21 will allow modulation of LEDs or laser diodes at a high frequency. It is also a two-stage amplifier with Q1 in a common-base configuration being the first stage, and Q2 and Q3 connected in quasi-darlington for the second amplifier stage. R4 and R5 form a voltage divider

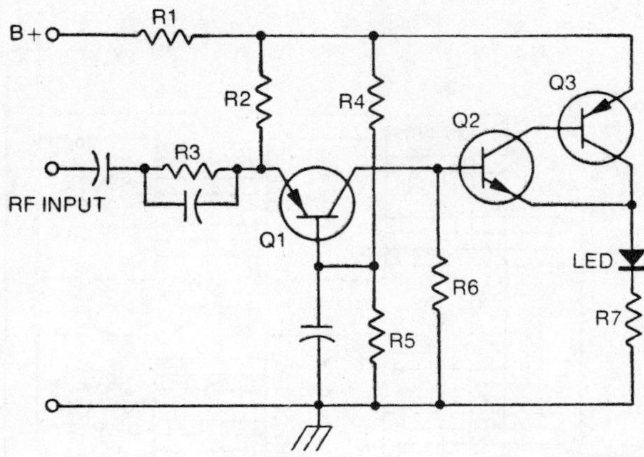

Fig. 5-21. With a circuit like this, an LED may be modulated at frequencies up to 100 MHz. This means that you could transmit all the AM broadcast stations and all the 27 MHz citizens' band channels on the single infrared light beam.

which sets the DC bias current on the LED via the transistor amplifiers.

DISPLAY CIRCUITS

The largest use of visible LEDs is in displays such as the seven-segment types. To drive these displays a decoder is required. In most every IC logic family there is at least one decoder-driver circuit suitable for the standard seven-segment displays. Figure 5-22 shows the use of the 7446 TTL, BCD to seven-segment decoder-driver. This device has a four-bit input line to receive the binary-coded decimal data and seven output terminals corresponding to the seven segments of the digit. This IC also has ripple blanking inputs, which allow the seven-segment displays to be multiplexed to reduce power-supply drain, and a lamp check input, which will energize all the segments independently of the data.

It is easy to see that with a system such as shown in Fig. 5-22, it would be impossible to build very small calculators and watches. In fact, most calculator and watch ICs have the decoders and multiplexers built directly in the main IC chip. In these applications, the segment and digit lines that are brought out are multiplexed together so that only one set of segment

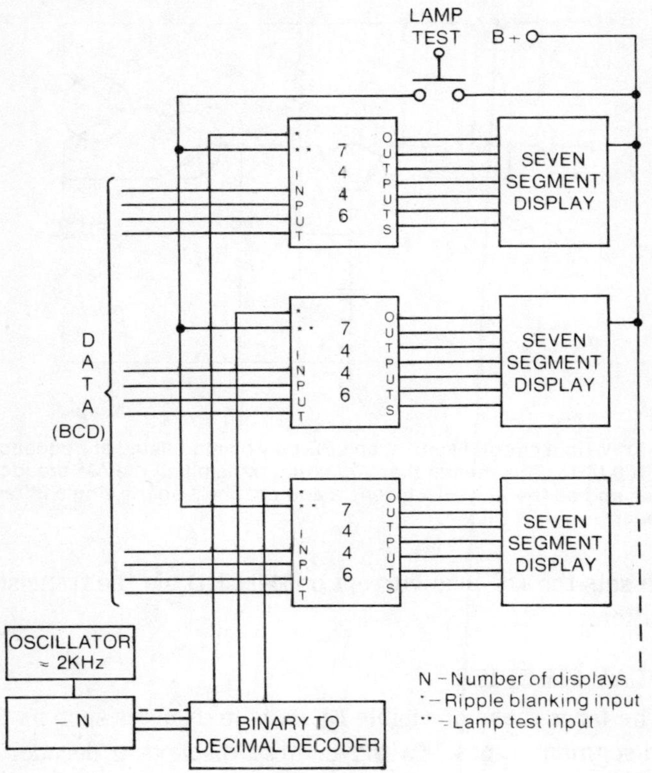

Fig. 5-22. When it is necessary to display binary-coded decimal data, a system such as this can be used. The 7446 BCD to seven-segment decoder drivers convert the data into the correct format for seven-segment displays such as LED displays. The use of the ripple blanking inputs allow multiplexing the displays.

lines plus one line for each digit is required, as shown in Fig. 5-23. The internal oscillator in the IC then sequences through the digits at two or three kilohertz, providing a display that appears continuous. In watch circuits, low-current displays are used which eliminate the need for the buffer transistors Q1 and Q2.

Driving a five by seven LED display requires a character generator. A unit such as the Signetics 2513 provides a 64 character output from the ASCII coded input. The method of interfacing this IC memory to a five by seven array is shown in

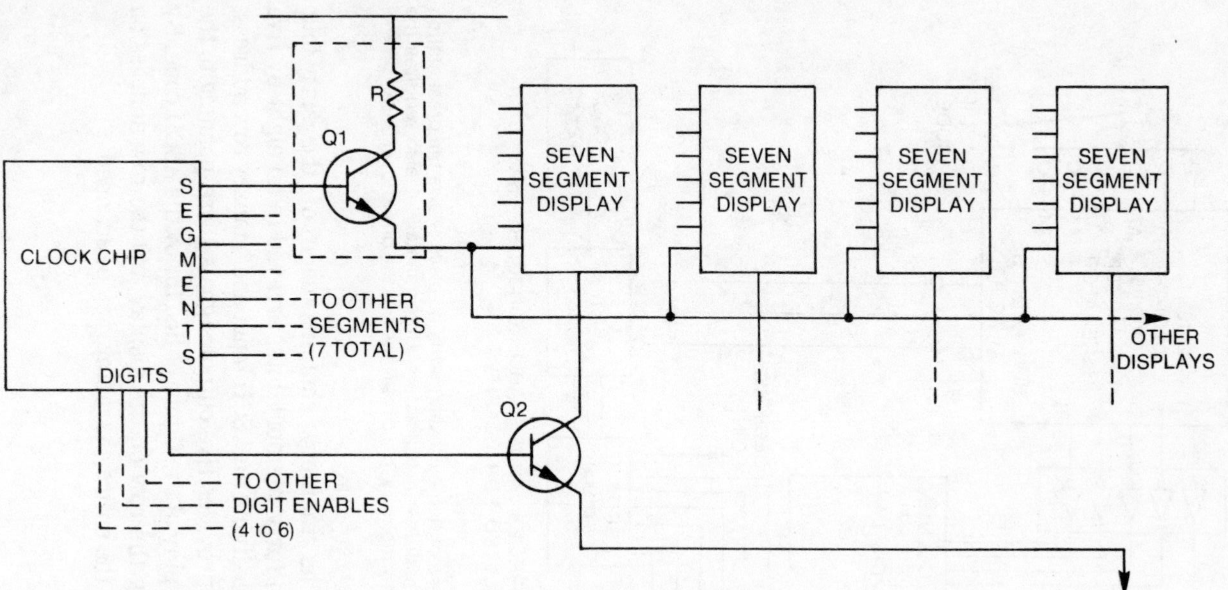

Fig. 5-23. Most IC clock chips have their own decoders and multiplexers built on the chip. This minimizes the number of pins necessary. For some miniature watch type LED displays the buffer transistors are not necessary.

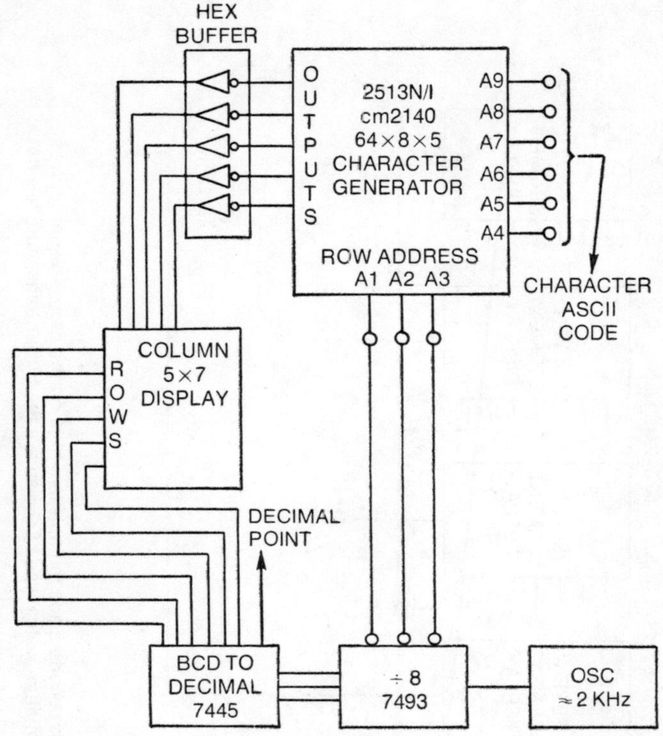

Fig. 5-24. The use of five by seven matrix displays allows the presentation of most all alphanumeric characters. Unfortunately, it also requires the use of a read only memory in a circuit such as the one shown here.

Fig. 5-24. The memory contains each of the sixty-four characters in forty memory bits, arrayed in an eight by five format. On receiving an ASCII input and a binary row address, the memory provides five output signals corresponding to the row of the display of the figure called for by the ASCII code. By synchronizing the row on the display and the row address on the memory, the entire character may be scanned.

Chapter 6 Troubleshooting Hints for Projects and General Electronic Circuits

The most frustrating thing that can happen (and most likely did or you probably would have skipped this chapter) is for a project or circuit not to work when first energized. The first thing to do is to take a coffee break. Electronics is not a black art, there is always a logical reason why an electronic circuit doesn't work. The task before us is to determine why.

GENERAL PROBLEMS

This section may get some of you upset, but I have had these problems myself and witnessed others with the same ones. The first thing is to determine if the circuit is getting power. Is it plugged in? Is the switch turned on? Are there blown fuses? For DC circuits, do you have the right polarity? It is surprising how often the answer is no to one or more of these questions. A VOM is a very handy tool for checking fuses. Fuses read at a very low resistance when okay or very high resistance when blown. After setting the VOM to the correct voltage range, the supply voltage can be checked. Be careful when working with line voltage; it can be dangerous.

A second category of general type problems are due to mechanical reasons. Loose, bad, or worn connections are the main culprits in this category. Are components in their sockets correctly? Are all the pins making good connection? Remember that when wiring components from the back of a circuit board the pins become reversed. More than one socket has been wired backwards. Now check for loose wires. The only wire that has only one end connected in electronic circuits is an antenna. Cold solder joints are very common in circuits built by novices and often sneak into even the most experienced circuit-builders circuits. If you are in doubt about the solder joint, resolder it.

If the obvious items suggested above do not correct the problem, then it will be necessary to dig into the system in detail to determine the problem. The main idea here is to do it logically. Every circuit, except the very simple, can be divided into simple functional blocks. Each block should operate in a given fashion. Once this is known, it is only necessary to check the inputs and outputs of these blocks to determine if the block is getting its input and producing its output. If the block does not have its input, then one of the blocks feeding it is causing the problem. If the output signal is okay, then the problem lies farther toward the output. If the input signal is okay but the output is wrong, then the problem is with the specific block under investigation. Each block can be examined in turn to locate the problems.

Once a fault is located in a functional block, it is then necessary to determine the cause of the fault. If we know what the block is supposed to do and what it actually does, then with a little thought it is possible to determine the problem.

The system described above is shown in Fig. 6-1. You can use the figure as a road map for troubleshooting.

A function block will usually contain less than six components. It may be necessary to check each of these to determine the problem. The following sections deal with each type of component, giving hints and techniques for testing each type.

RESISTORS

As a rule resistors are the most reliable components used in electronics. This holds especially for carbon composition types. The resistor used in a circuit should be chosen for

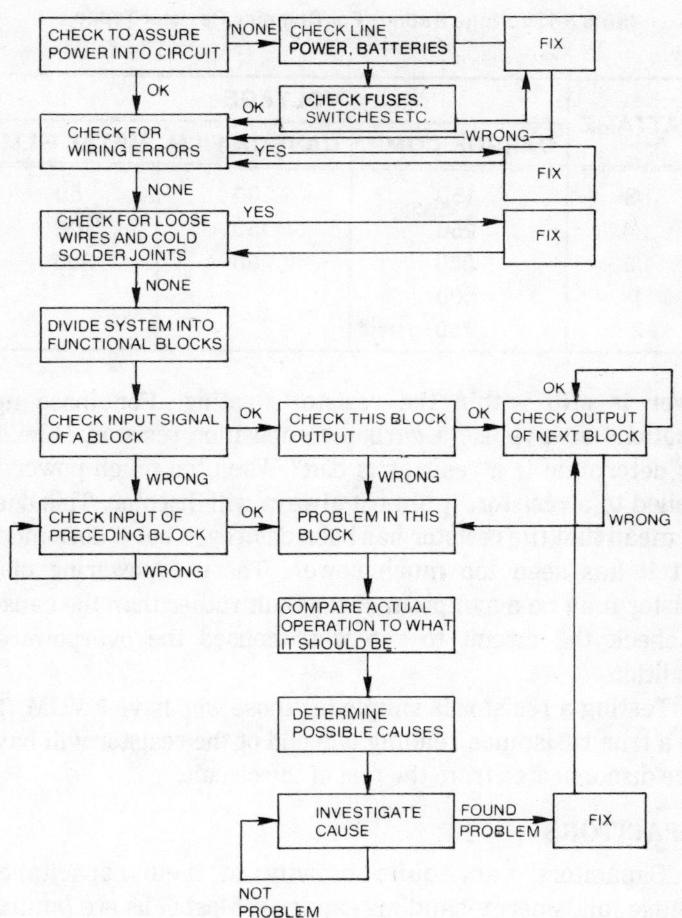

Fig. 6-1. The secret to troubleshooting is to proceed systematically. If circuit doesn't work, start at the top of the chart and follow through the paths. When you fix something it may be necessary to back up a bit.

application. Each resistor is characterized by its resistance, a tolerance, a power rating, and a voltage rating. Many people are unaware of the last and run into problems due to arcing in the resistor. Table 6-1 gives the voltage ratings for the three most popular types of resistors. If you aren't sure what voltage the resistor will see, assume the worst and go from there. In some applications a resistor may see pulses of energy much higher than its power ratings, but the average

Table 6-1. Voltage Ratings For Common Resistor Types.

WATTAGE	VOLTAGE		
	CARBON COMP.	CARBON FILM	METAL FILM
1/8	150	100	50
1/4	250	150	100
1/2	350	250	200
1	500		
2	750		

power is still within the resistor's rating. For these applications always use a carbon composition resistor. How do you determine if a resistor is bad? When too much power is applied to a resistor, it almost always will discolor. This does not mean that the resistor has been destroyed but it does mean that it has seen too much power. The overpowering of a resistor may be a symptom of the fault rather than the cause, so check the circuit to see what caused the overpowered condition.

Testing a resistor is simple for those who have a VOM. To get a true resistance reading one end of the resistor will have to be disconnected from the rest of the circuit.

CAPACITORS

Capacitors vary quite broadly in their capacitance, voltage, and energy-handling capacity. Most of us are familiar with electrolytic and ceramic capacitors, but in many applications it is necessary to use film, mica, tantalum, or one of the many other types of capacitors. Each has its own characteristics and may not be suitable for some applications. Table 6-2 gives a breakdown of the major capacitor families and their main applications.

In a circuit, remember that tantalum and electrolytic capacitors are used with DC bias voltages and that the capacitors themselves are polarized. If inserted in the wrong polarity they may be damaged. Electrolytic capacitors wired into power supplies in the wrong polarity have been known to explode, so be sure to wire these in at the correct polarity.

Table 6-2. Various Types of Capacitors For Electronic Circuits.

TYPE	CAPACITANCE RANGE	VOLTAGE RANGE	MAIN APPLICATIONS
Mica	10pF-0.01μF	50-1000V	high frequency (RF) high energy/μF, high Q at high frequency.
Ceramic	10pF-0.5μF	50-5KV	medium frequency, lower energy
Film	0.001μF-100μF	50-1000V	medium energy, low frequency to RF, AC voltage
Foil	0.001μF-100μF	25-5KV	higher energy than Film, low frequency to RF, AC voltage
Electrolytic	0.5μF-10,000μF	5-1KV	polarized, energy storage, power supplies, etc.
Tantalum	0.1μF-1000μF	5-200V	most polarized, low leakage, timing

Capacitors may be checked for shorts with an ohmmeter, and should indicate high resistance. On many electrolytic and tantalum capacitors the resistance is lower in the reverse polarity.

Checking the capacitance of a capacitor generally requires more instrumentation than is available to the experimenter. A simple circuit such as shown in Fig. 6-2 can be used to get a good idea of the capacitance of a capacitor. When used in conjunction with a VTVM, an unknown capacitor can easily be matched to a known capacitor. To use the circuit, the correct values of resistors are inserted into the circuit along with a known good capacitor (C_K) of a similar capacitance. With a VTVM on AC volts, measure the voltages from A to C and from B to C. The value of the unknown capacitor (C_U) will be approximately:

$$C_U = C_K (V_{A\text{-}C} / V_{B\text{-}C})$$

Even for a first estimate, the formula will yield reasonably close answers. For example, suppose the unknown capacitor is exactly 0.05 μF and you select a known capacitor of 0.1 μF. Using two 100K resistors in the test circuit, you measure the voltage and find $V_{A\text{-}C}$ is 1.61 volts and $V_{B\text{-}C}$ is 2.74 volts. The computed value is then $0.1 \times 1.61/2.74 = 0.058$ μF, which is fairly close. Trying a known capacitor of 0.05 μF in

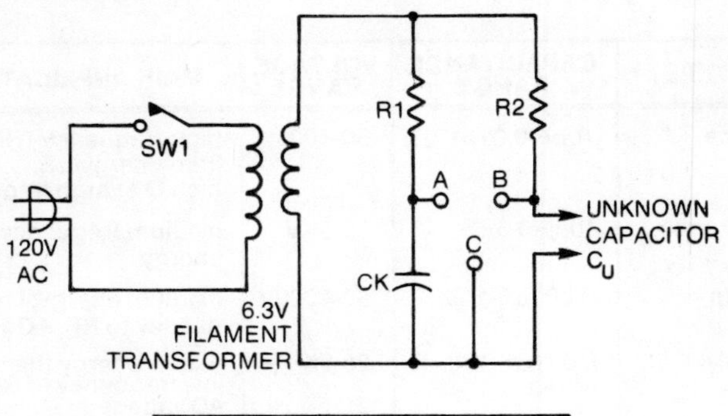

APPROXIMATE VALUE OF UNKNOWN CAPACITOR	R1 and R2
100 – 1000 pF	10 meg
1000 – 10,000 pF	1 meg
0.01 – 0.1 μF	100K
0.1 – 1.0 μF	10K
1.0 – 10 μF	1K
10 – 100 μF	100Ω

Fig. 6-2. A simple capacitance comparison bridge. For better results for low capacitance values a higher frequency signal should be used. As the frequency is increased, R1 and R2 should be reduced by the ratio of the new frequency to 60 Hz.

the circuit would then yield a much more accurate value of the unknown capacitor.

A VOM can be used for capacitances as low as 1.0 microfarad with the circuit as shown. For smaller capacitors a VOM can be used with a higher frequency source in place of the filament transformer. For those who enjoy building their own equipment a multipole rotary switch can be used to assemble the components for the different ranges and a battery-operated oscillator can be used for the signal source.

TRANSFORMERS AND INDUCTORS

Transformers and inductors generally cause few problems. They come in many shapes and sizes, but both classes have many turns of wire which has usually been wrapped around a form or a magnetic material. Inductors usually contain one wire that runs from one terminal or lead to the other. Transformers, on the other hand, generally have two or more isolated windings. Both often have taps or connections along the windings, giving rise to even more terminals.

The main problem that occurs with transformers in building circuits is that they often are wired wrong. With an ohmmeter it is possible to find all the terminals of a winding merely by measuring the resistance between one lead and another. A low impedance indicates a common winding between leads, while a very high or open indicates different windings. On a transformer, the manufacturers never use wire larger than required. This in terms of current means that the larger wires are the low-voltage windings, while the finer wires are the high-voltage windings.

Transformers and inductors when applied properly are very reliable. However, other faults in a system or miswiring may damage them. This generally occurs when too high of a current is conducted through the windings. This causes overheating which can cause insulations to break down, giving shorted turns in a coil or between adjacent windings. Shorts between windings, or between windings and the frame, can be found with an ohmmeter, while shorts between turns of the same winding are more difficult to find. To check for shorted turns, rated voltage is applied to each winding with all other windings open. Shorted turns will usually appear as excessive current through the winding. If a second transformer of the same type is available, the currents can be compared.

VACUUM TUBES

Of all the components in electronics, tubes cause by far the most problems. For this reason most modern electronics are built with only semiconductor devices where possible. The most common problem with tubes is burned out filaments. A simple ohmmeter check across the proper pins will give a good indication of this problem. In some circuits the filaments are wired in series. For these, an in-circuit check with a voltmeter can find the open filament, since the entire string voltage will appear across the open device.

As a rule, if the equipment is tube operated check all of the tubes. In 90% of the cases the problem will be a bad tube.

SEMICONDUCTOR DEVICES

Semiconductor devices are, in general, very reliable but they can be damaged by handling and improper application. When handling these be careful of the leads of the device. It is often necessary to bend or cut the leads. It is important not to transmit stress from the lead to the package. When bending or cutting the leads, hold the lead firmly between the device body and the bend or cut. Secondly, never try to modify a package by cutting off the mounting stud or mounting flange. This will undoubtedly cause inside damage to the device. Many devices are also very sensitive to overheating during soldering. This is especially true of germanium transistors and diodes. When soldering a device into a circuit, a small metal clip should be attached to the lead between the solder joint and the device body to prevent the heat from being transmitted to the device.

Heat is one of the biggest problems with semiconductor power devices. Power transistors, rectifiers, and thyristors very often require a heat sink. As a rule of thumb, if you can't keep your finger on the device, it's too hot and therefore needs a heat sink (or more heat sinking). Remember, the cooler the semiconductor runs the better its reliability. A problem that often crops up with heat sinking is insulating the device or heat sink from the rest of the circuit. Careful checks with an ohmmeter will reveal any short circuits. Remember to check all device leads as well as the package for shorts.

Bad semiconductor devices are sometimes difficult to locate. A VOM is a handy instrument for finding gross problems such as shorts or opens, but it is not capable of finding degraded or out-of-specification parts. A diode junction will show low resistance in the forward direction and high resistance in the reverse direction on an ohmmeter. Since most devices used are junction devices, the ohmmeter gives the information that the junction is there and that there are connections to the junction; assuming anything else from the ohmmeter reading is foolish. In circuit voltage checks they are also useful, since they indicate the polarity and magnitude of the voltages. Before attempting to interpret voltmeter or ohmmeter readings, be sure how the instrument operates and what the meter indicates. Techniques for each of the semiconductors will be discussed.

DIODES AND RECTIFIERS

Though many use these names interchangeably, a rectifier is only one type of diode. A diode can be any two-terminal semiconductor or tube-type device. For the purposes of this book it is impractical to cover all the types of diodes, so we shall only consider rectifiers and zener diodes. For both types, out-of-circuit measurement with an ohmmeter

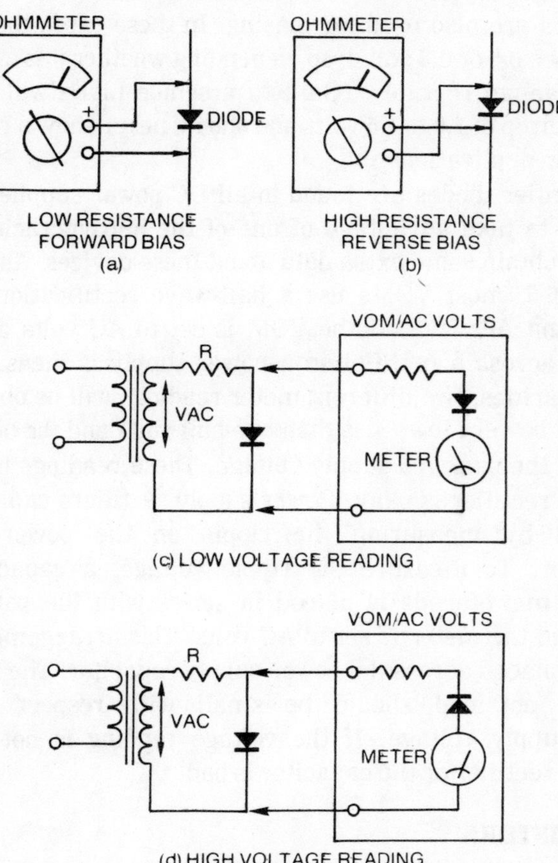

Fig. 6-3. Using a VOM to check rectifier diodes. (a) and (b) shows an out-of-circuit ohmmeter test. When the diode is forward biased as in (a), the meter will indicate lower resistance than when reverse biased as in (b.) (c) and (d) show in-circuit voltmeter readings. Remember VOMs use rectified voltage for AC measurements and therefore only see half-wave voltage. If the meter sees the forward drop it will show a low voltage as in (c). If the meter sees the reverse voltage it will show the entire supply voltage.

shows a low impedance in one direction and a higher impedance in the other.

In the circuit, voltmeter readings give a good indication of whether the device is functioning properly. Zener diodes are usually to set voltage bias levels. Therefore, in most applications, a voltmeter reading across a zener diode will indicate the zener voltage. In all other circuits the voltage reading across the zener will be less than its zener voltage. Rectifiers are also used for biasing. In these cases the diode will show a 0.6 or 0.3 volt drop, depending whether it is a silicon or germanium rectifier. (Gallium arsenide LEDs will give a forward drop of 1.0 to 1.5 volts and should never show a reverse voltage in applications.)

Rectifier diodes are found in all DC power supplies. It is possible to take advantage of one of the idiosyncracies of a VOM to obtain some extra data about these devices. As shown in Fig. 6-3, most VOMs use a half-wave rectification when measuring AC volts. If the VOM is set to AC volts and the voltage across a rectifier in a power supply is measured in both polarities, two different meter readings will be obtained. One will be very low (less than one-half volt) and the other up to twice the peak AC supply voltage. These readings indicate that the rectifier is okay. Power-supply rectifiers can also be checked by measuring the ripple on the power-supply capacitor. To measure the ripple voltage, a capacitor of several microfarads is placed in series with the voltmeter leads and the meter is set to AC volts. This arrangement can then be placed across the power-supply capacitor. The voltage reading obtained should be small with respect to the power-supply voltage. If the voltage reading is not small, either a rectifier or the capacitor is bad.

TRANSISTORS

When transistors are used in circuits, problems are often encountered by getting the leads reversed. Figure 6-4 shows that a transistor can be considered as two diodes, one a zener and the other a normal rectifier diode. When a VOM (used as an ohmmeter) is connected to the transistor it will show high impedance in both directions if it is connected col-

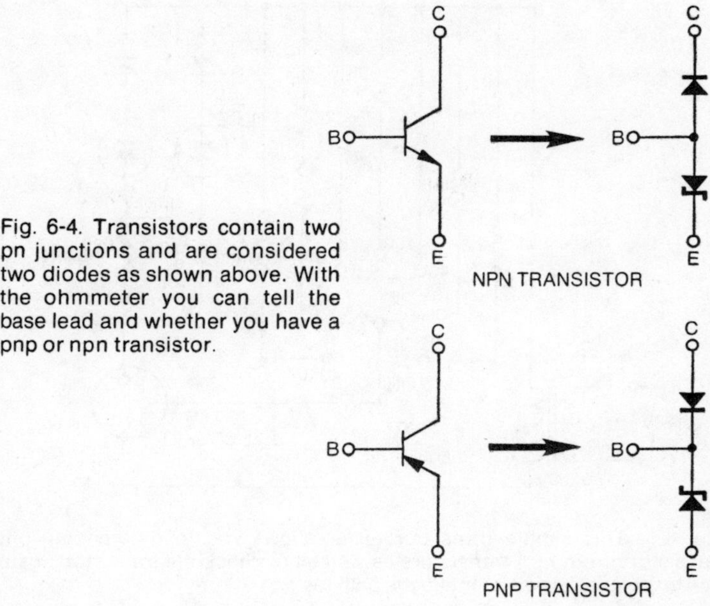

Fig. 6-4. Transistors contain two pn junctions and are considered two diodes as shown above. With the ohmmeter you can tell the base lead and whether you have a pnp or npn transistor.

lector-to-emitter. If connected from base to either collector or emitter it will show a diode. After checking to find which lead is positive, a pnp can be separated from an npn transistor.

The transistor tester shown in Fig. 6-5 provides a simple method of checking transistors. Although it is not the most exotic piece of equipment that you will ever build, it will allow you to distinguish a pnp from an npn transistor as well as getting a rough measurement of the transistor's gain. To measure the gain, the rotary switch is rotated from the low settings to the higher until the appropriate LED just begins to dim.

SCRs AND TRIACS

Thyristors, such as SCRs and triacs, are much more difficult to test than transistors and rectifiers. In out-of-circuit tests, the gate can be found with an ohmmeter. The equivalent diode circuits of SCRs and triacs are shown in Fig. 6-6. Other tests on SCRs and triacs require high-voltage circuitry. The easiest way to test for blocking voltage and the like is with a voltmeter in the circuit. When the SCR or triac is in the *off*

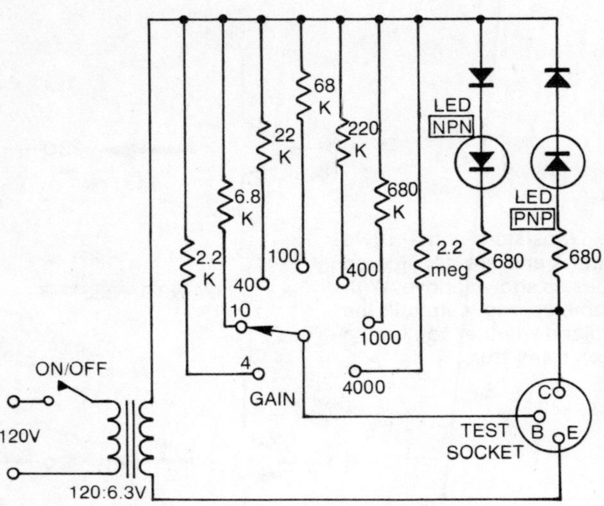

Fig. 6-5. This simple transistor tester allows you to differentiate pnp transistors from npn transistors as well as to check the transistor's gain. The text explains how to make the gain test.

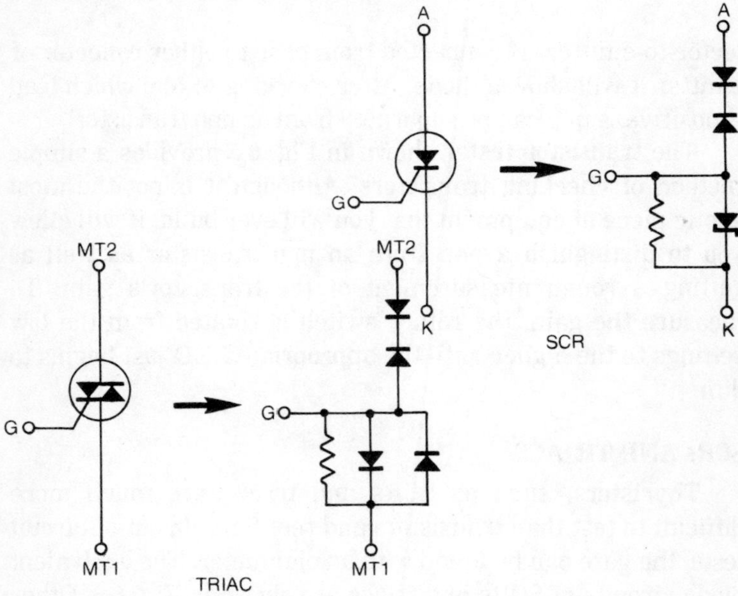

Fig. 6-6. Triacs and SCRs can also be considered series diodes like transistors. The resistors shown are not in sensitive gate devices. Note the anode and MT2 terminals will appear as high impedances to all other terminals regardless of meter polarity.

state the full supply voltage will be seen across the SCR. In the *on* state an SCR will appear like a rectifier, and the triac will appear like two rectifiers connected in reverse parallel. VOM voltage checks should be made in both directions for both devices.

One problem that continually crops up with SCR and triac currents is inadequate heat sinking of the power devices. If there is a mounting stand or hole, use it. Again, if you cannot hold your finger on the package, it is too hot and needs more heat sinking.

Projects

1

Two Solar-Powered Radios

One of the most fascinating, simple, and inexpensive projects one can build is a radio which derives its power from the sun. The circuit shown in Fig. P1-1 should only cost about five dollars, while the two-stage circuit shown in Fig. P1-2 would run around eight dollars. If you live in an area where radio reception is not very good, it may be necessary to build the two-stage circuit to get decent reception.

The two circuits operate in similar fashion. The signal from the radio station is received by the antenna. The coil and tuning capacitor form a filter which allows the tuned frequency to pass to the diode. The diode acts as the demodulator and converts the radio frequency back to audio frequency. This is the inverse of what occurred at the radio station. The transistor then amplifies the audio signal to drive the headphone in the single-stage radio or the interstage transformer in the two-stage radio. The second stage of the two-stage radio serves only to amplify the sound.

Either of these circuits can be constructed in any small box that is about the size of a small transistor radio. For the Scotsman, ask at your local pharmacy for several of the plastic boxes in which ointments are delivered. They are in handy sizes and the pharmacist usually throws them away anyway. The author uses these free containers for holding all

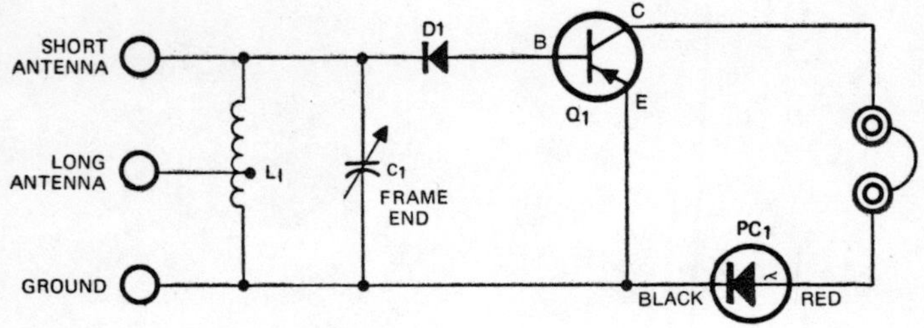

SOLAR RADIO – SINGLE STAGE – PARTS LIST

Q1	PNP Transistor (IR-TR05 or TR14)
C1	365pf variable Capacitor (J.W. Miller No. 2111, or equivalent)
L1	Transistor Loopstick Antenna (J.W. Miller No. 2001, or equivalent)
D1	Germanium Diode (IR-1N34A)
PC1	Selenium or Silicon Solar Cell (IR-B2M, B3M, or S1M)
HEADPHONES	2,000 or 4,000 Ohms
OPTIONAL:	Transistor Socket (IR-SH130 or SH131)
	Printed Circuit Board (IR-PCB21 or PCB.3)
	PC Connector Path (IR-CS60)
	PC Patterns (IR-CZ30, CZ50, CZ71)

Fig. P1-1. Single-stage solar-powered radio schematic and parts list. (Courtesy of International Rectifier.)

sorts of miniature parts, in addition to using them to build projects. Both radios are designed for either a short (10–20 ft) or a long (20–50 ft) antenna. The radio will probably work best with the long antenna, but you can get very good reception by experimenting with the position of the short antenna.

When the coil is examined, three wires will be seen. The one that is doubled up (that is two wires together) is the long-antenna connection. The lead nearest to this is the ground lead. The remaining lead is the short-antenna connection which gets connected to the lugs on the tuning capacitor. Connections to the frame of the tuning capacitor should be made by inserting a small screw into one of the front holes and wrapping the wire under the screw head. Be careful that the screw does not touch the aluminum plates of the capacitor.

The diode is connected with banded end to the capacitor lug. The transistor leads are identified in Fig. P1-1. The photovoltaic cell should be mounted so that maximum light will shine on it. For increasing the volume a second cell may be added. It should be connected in series with the first cell such that the red lead of one cell is connected to the black of the other. The two remaining leads, the black of one and the red of the other, are then attached as shown in the figures.

The output power of the one-stage transistor radio can be increased by adding a second transistor stage as shown in Fig. P1-2. A transformer is used to couple the output of the first stage into the input of the second stage. The transformer may be any interstage-type rated at 10,000 ohms to 2,000 ohms. The lead colors refer to a Stancor TA-35. For other transformers, refer to the instruction sheet (which is usually enclosed with the transformer) to determine the proper lead colors.

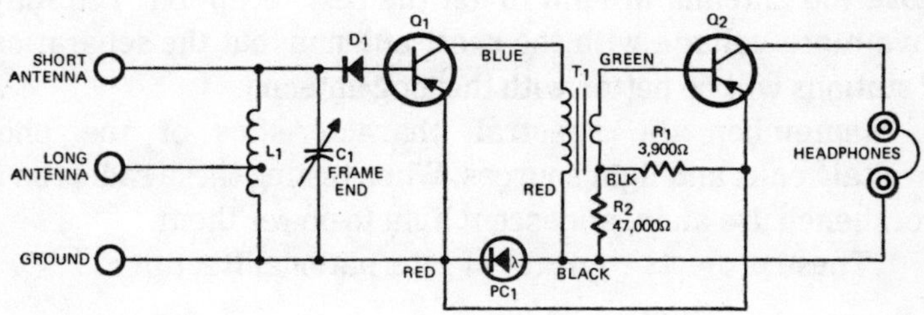

SOLAR RADIO – TWO STAGE – PARTS LIST

Q_1, Q_2	PNP Transistors (IR-TR05 or TR14)
PC_1	Selenium or Silicon Solar Cell (IR-B2M, B3M, or S1M)
T_1	10,000 Ohm to 2,000 Ohm Interstage Transformer (Stancor TA-35, or equivalent)
L_1	Transistor Loopstick Antenna (J.W. Miller No. 2001, or equivalent)
C_1	365pf Variable Capacitor (J.W. Miller No. 2111, or equivalent)
R_1	3,900 Ohm Resistor
R_2	47K Ohm Resistor
RD_1	Germanium Diode (IR-1N34A)
Headphones	
	Transistor Socket (IR-SH130 or SH131)
	Printed Circuit Board (IR-PCB21 or PCB23)
	PC Connector Path (IR-CZ60)
	PC Patterns (IR-CZ50, CZ30, CZ70)

Fig. P1-2. Two-stage solar-powered radio schematic and parts list. (Courtesy of International Rectifier.)

One problem which can occur with the two-stage radio is that there may be too much volume for clear reception. This may be easily corrected by placing a 100 μF low-voltage capacitor in parallel to the solar cell. This will allow much higher peak currents to be drawn by the radio and should clear up the reception.

For the best reception, connect the receiver to an antenna of 20 feet or more. The longer the antenna the greater the volume and the more stations received. In large cities, the long antenna may create problems of station-to-station interference. A good ground to a cold-water pipe or a four-foot metal ground stake driven into the ground will improve the selectivity of the radio.

When testing the radio, try both antenna connections and move the antenna around to get the best reception. You may have more volume with the short antenna, but the separation of stations will be better with the long antenna.

Remember the spectral characteristics of the photovoltaic cells and light sources. When testing these radios on a workbench use an incandescent light to power them.

(These projects courtesy of International Rectifier)

2

Light-Beam Transmitter and Receiver

In several large cities in the U.S., optical data links are used to interconnect two or more computers at different locations. These systems send beams of infrared light from rooftop to rooftop, carrying computer data that would require several high-frequency phone lines for each light beam. These systems cost many thousands of dollars but a very simple transmitter/receiver can be made in a few hours with some very inexpensive components.

Figure P2-1 shows the schematic of the light-beam transmitter. To keep it simple, an ordinary flashlight bulb is used as the light source and a carbon microphone to pick up the audio signal. The single transistor serves to amplify the audio signal, and the transformer T2 modulates the current of the lamp. Battery B1 provides the bias for the microphone while B3 biases the lamp. Battery B2 powers the single transistor amplifier. The assembly of the circuit is straightforward. The high-voltage windings of the two transformers are connected to the transistor and the low-voltage sides connected to the microphone and lamp circuits. The transformers can be of any type that has the turns ratio of about 20 to 1. The transformer listed in the parts list has been tried and works well, but a Stancor A-3856 gives a

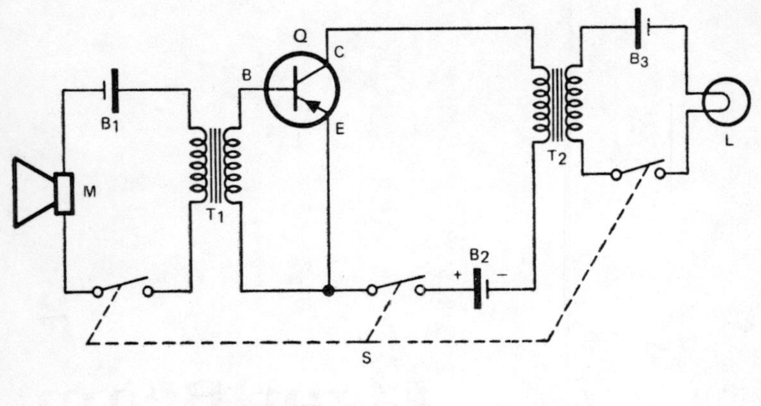

PARTS LIST

M = Single button carbon microphone, available through surplus outlets
T_1 = T_2 = Transformer (Triad F-13X sec. 6.3V @ 0.6A, pri. 115V)
Q = PNP transistor (IR-TR-05)
L = Flashlight lamp 112
S = 4-pole single throw switch (Mallory type 3242 or 744)
B_1 = 7.5V battery (Eveready 717, Burgess 5540, or equivalent)
B_2 = 22.5V battery (Burgess V15 or equivalent)
B_3 = Size D flashlight cell

Fig. P2-1. Light-beam transmitter schematic and parts list. (Courtesy of International Rectifier.)

little better performance. Most electrical experimenters have a few 6.3V filament transformers lying around and a pair of these should prove adequate. The switch is important to conserve battery life and can be any three- or four-pole switch that can switch 250 milliamps.

The type 112 lamp was chosen because of its low-mass filament. Due to the small filament it can respond to frequencies up to two to three thousand hertz. This is more than adequate for voice transmission. A second added feature of the 112 lamp is its built-in lens at the top of the bulb. This allows the majority of its light output to be concentrated in a small area. For distances of about twelve feet, two lenses should be used with the lamp. The first, placed near the lamp, should be a three-inch diameter, four-inch focal-length convex lens. The separation of the lens and the lamp should be adjusted for maximum light intensity on the lens of the receiver. The lens at the receiver should be a six-inch or larger convex lens with a focal length of eight to twelve inches. The lens-to-cell distance should be such that the light is concentrated on the total active area of the photovoltaic cell. For minimal cost, these lenses should be single planoconvex condensing lenses, though any lens meeting the criteria above will be adequate.

The light-beam receiver circuitry consists of the photovoltaic cell with a two-stage direct-coupled amplifier. (See Fig. P2-2). The headphones are driven through an impedance-matching transformer. The output transformer can be any impedance-matching transformer for the headphones used which has a primary impedance of 15K to 25K. The types listed in the parts list will suffice for most high-impedance headphones. For a more miniaturized receiver, either a U.T.C. SS03 or a Stancor VM-110 can be used. It is also possible to drive a small permanent-magnet speaker with the receiver. A Stancor A-3857 or Thordarson TS24554 will adequately drive a small low-impedance speaker.

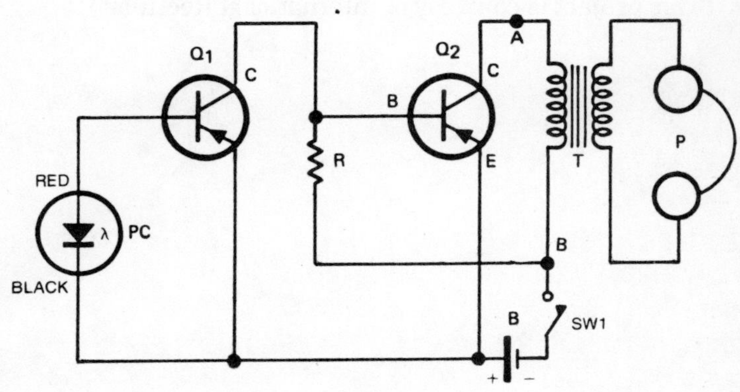

PARTS LIST
PC = Photocell (IR-B2M)
Q1 = Q2 = PNP Transistor (IR-TR-05)
P = Headphones (see text)
SW1 = SPST switch
R = 330,000 ohm, ½ watt resistor
B = 2.5V battery (Burgess U15)
T = Stancor A-3250, Triad-T-23, or Thordarson T22590

Fig. P2-2. Light-beam receiver schematic and parts list. (Courtesy of International Rectifier.)

To test the transmitter, turn the unit on and see if the lamp flickers with the sound. If it does, this indicates the transmitter is working properly. With the lenses removed, place the transmitter and receiver several feet apart with the light beam aimed into the photovoltaic cell. The active area of the photovoltaic cell should be perpendicular to the light beam to receive the maximum light. With a VOM on AC volts across the headphone terminals, a deflection should be seen when speech or music is directed into the microphone. The lenses and alignment of the light path can be adjusted for maximum output by replacing the battery and microphone with an AC signal generator. Signals of 500 Hz to 2 kHz will work best.

For those more skilled, ranges of hundreds of yards can be achieved using high-powered LEDs or semiconductor laser diodes. Since these devices operate in the infrared region, they will produce no visible light but will allow for much higher energy to be transmitted. Extra caution must be observed with this modification, since the high energy can cause permanent damage to the eyes. The 112 lamp provides a lower cost and very safe way to experiment with light-beam transmission.

(This project is courtesy of International Rectifier.)

3

Two Slave Flashes

Here are two slave flash circuits for the amateur photographer designed to trigger when the camera-connected flash is triggered. The first is a simple unit which can be used with most types of flashbulbs. The second is a xenon flash unit.

The light-triggered flashbulb unit is a relatively simple capacitor-discharge circuit. The circuit is shown in Fig. P3-1. The supply voltage for the circuit is from either a 22.5V battery or two nine-volt radio batteries wired in series. A capacitor is charged to battery voltage through a current limit resistor R1. When a bright light falls on the light-activated SCR, the SCR turns on and discharges the capacitor through the flashbulb. The flashbulb, before it is flashed, is of very low impedance so when the SCR triggers, the filament conducts a large current. As the lamp filament burns, the lamp becomes a very high impedance. This stops the current through the SCR.

The circuit components with the battery and switch can be easily mounted inside of a minibox. The SCR should be mounted into a side panel of the minibox so that the light from the main flash can turn it on. The flashbulb socket and reflector, if used, may be mounted to the top of the minibox. The slave flash reflector should be mounted so that it may be

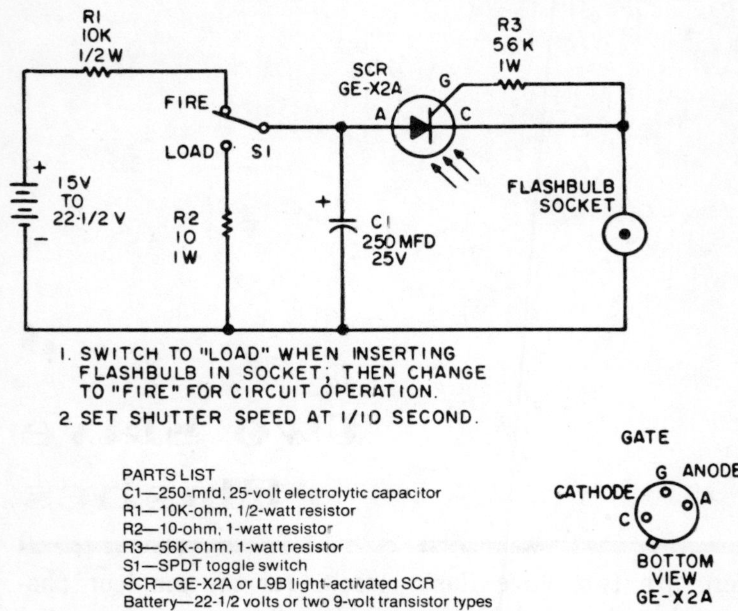

Fig. P3-1. Flashbulb, remotely triggered flash. (Courtesy of General Electric Semiconductor Prod. Dept.)

rotated to allow the flash to be directed, while allowing the SCR to be aimed toward the camera.

The second slave flash is shown in Fig. P3-2. This 50-watt-second slave flash uses a xenon flashtube for the flash element, allowing 1/1000 of a second synchronization between the main flash and the slave. With flashbulbs the delay time can be as great as 1/10 of a second. The circuit is typical of xenon flashes. The power is supplied by either a high-voltage battery pack or by the voltage-doubler circuit. For those not familiar with a voltage doubler, the two rectifiers each conduct on alternate half-cycles. When one of the rectifiers conducts, it charges the capacitor to the peak output voltage of the transformer (in this case, 170V). The voltages of the two capacitors are in series so that across points A and B the voltage is twice the peak supply voltage. This voltage supply or the optional battery pack, which can be connected instead to points A and B, provide the high voltage to charge the 1000 microfarad energy-storage capacitor. It is

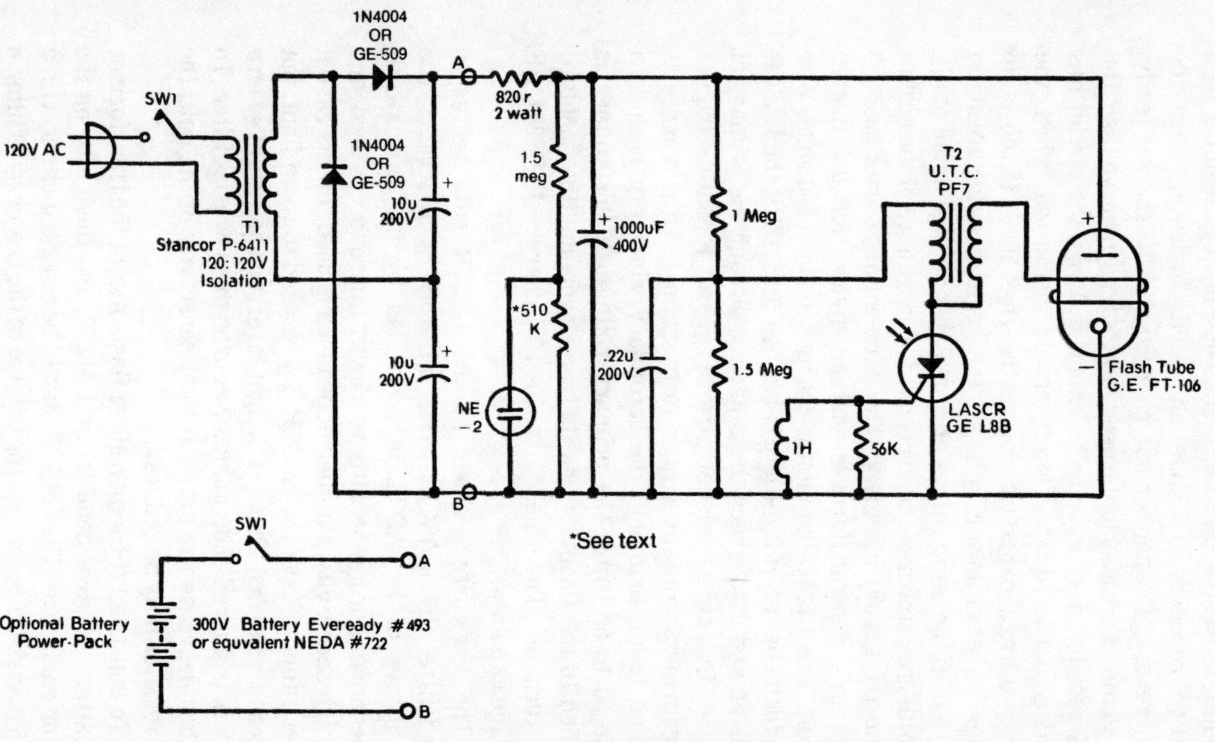

Fig. P3-2. Xenon slave flash circuit.

the energy from this capacitor which will be discharged through the xenon flash tube. The NE-2 neon glow lamp, with the two resistors, 1.5 meg and 510K, indicates when the energy-storage capacitor has charged and the flash is ready. Due to the variations in the neon lamp's triggering voltage, the 510K resistor may have to be changed. When the circuit has been constructed the lamp should come on when the energy-storage capacitor voltage reaches 300V. If the lamp comes on at a lower voltage, the resistor value must be reduced. If the lamp comes on at a higher voltage or not at all, the 510K resistor must be increased. The balance of the circuit is the triggering circuitry. The 1 megohm resistor and 1.5 megohm resistor form a voltage divider for the trigger circuit. The 0.22 μF capacitor charges to about 200 volts. Whe a sudden change of light level hits the L8B LASCR the LASCR turns on and discharges the capacitor through the primary of T2, a flashtube trigger transformer. Flashtube-trigger transformers contain a high turns ratio. When a pulse is applied to its primary, the secondary winding generates a pulse 30 to 50 times the primary voltage. If this voltage is applied to the flashtube via its trigger winding, the flashtube will turn on. The flash tube then discharges the 1000 μF storage capacitor.

The low-current choke on the LASCR gate serves to desensitize the LASCR from ambient light levels. The sensitivity of a LASCR is controlled by its gate-cathode impedance. For low sensitivity a lower impedance is used, and for high sensitivity a higher impedance is used. In this case it is desirable to have the LASCR not respond to room light, but be sensitive to fast changes in light level such as occurs when a flash is triggered. The inductor acts as this low-pass filter. To further decrease the ambient light sensitivity of the unit, the 56K resistor may be reduced.

To construct the xenon slave flash, mount the transformer or battery, switch, neon lamp, LASCR, and flash tube on the box or enclosure. The LASCR should be mounted either with a TO-5 transistor socket on the outside of the box or by drilling a hole large enough for a grommet which will hold the LASCR snugly. A drop of epoxy glue between the LASCR and grommet will hold it securely. The 1H choke and 56K resistor should be mounted on the SCR socket or LASCR leads as closely to the LASCR as possible. The balance of the components can be mounted on a piece of perforated board. Care should be taken on several items. The polarities of the 10 μF and 100 μF capacitors must be observed. The trigger transformer's output voltage is quite high, it is therefore important to keep the trigger-to-flashtube wire away from any of the circuit or the case.

4
Automatic Night-Light

When the house is left unattended, when you are away for an evening or vacation, or if you would like the porch light on when you return at night, this little circuit will come in very handy. With it you can automatically turn on up to 500W of lamps when it becomes dark.

The circuit is built around a photocell and reed-switch combination (see Fig. P4-1). The circuit operation can be explained as follows. The diode CR1, resistor R1, and capacitor C1 make up a DC power supply which is about 160V. The components in parallel with C1 form a voltage divider. The voltage is divided between R2 and the rest of the components in the network. When the photocell has little or no light it is a very high resistance. This allows the voltage across the reed switch to be great enough for it to close the circuit. When the photocell is exposed to light, its impedance drops. The result, low resistance, deprives the reed-switch coil of sufficient current to hold the reed switch closed. The closure of the reed switch energizes the gate circuit of the triac which acts as the power switch. The turning on of the triac allows current to flow in the lamps attached to the load receptacle.

Wiring of the components is not critical. The components can be assembled on perforated board or with two four-point

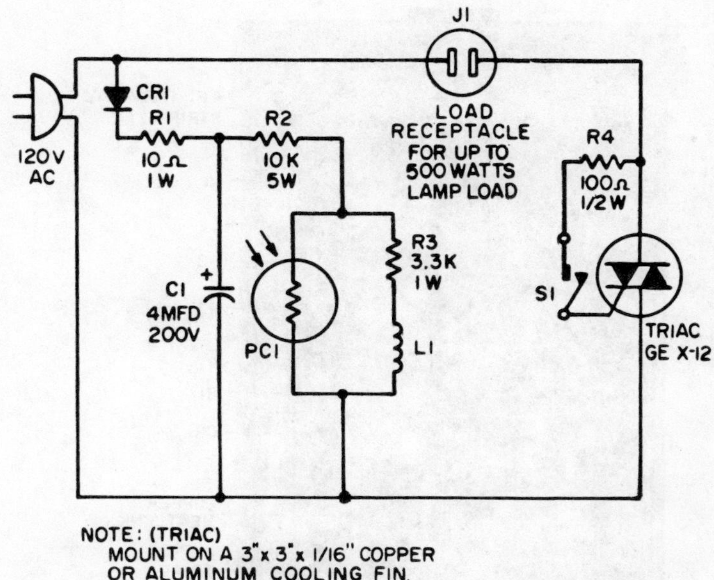

PARTS LIST

C1—4-mfd, 200-volt electrolytic capacitor
CR1—GE-504A rectifier diode
J1—Power socket: Amphenol Type 61F, or equivalent
L1—Reed switch coil: 10,000 turns of No. 39 enameled wire on ¼′ form 2″ long. This is coil C-2 at GE distributors.
PC1—GE-X6 phtoconductive cell
R1—10-ohm, 1-watt resistor
R2—10,000-ohm, 5-watt resistor (or two 22,000-ohm, 2-watt resistors in parallel)
R3—3,300-ohm, 1-watt resistor
R4—100-ohm, ½-watt resistor
S1—GE-X7 reed switch Triac—GE-X12
Minibox—5′ × 2¼′ × 2¼″; Bud CU-2104A, or equivalent
Misc.—Line cord, cable clamp, grommet (or epoxy), mounting screws, wire and solder, heatsink

Fig. P4-1. Automatic night-light parts list and schematic. (Courtesy of General Electric Semiconductor Products Dept.)

terminal strips as shown in Fig. P4-2. The photocell can be mounted with a grommet through one end of the minibox. A dab of epoxy will hold it securely in place. The load receptable can be mounted either on the opposite end or on the top of the box as shown in Fig. P4-2. Since resistor R2 dissipates

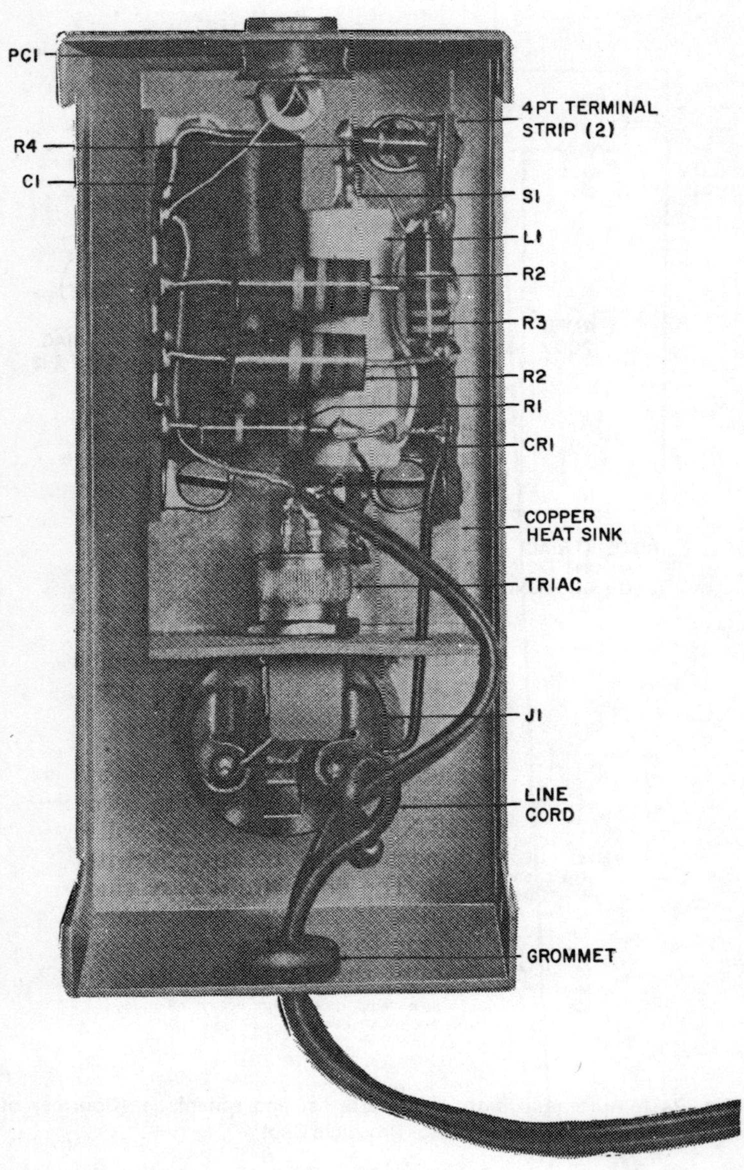

Fig. P4-2. A typical layout for the 500-watt version of the automatic night light. (Courtesy of General Electric Semiconductor Products Dept.)

considerable heat, it should be mounted away from the photocell. Caution should be taken that no electrical connections touch the box. The triac heat sink also must be kept electrically isolated from the box. This can be accomplished with either electrical tape or nylon screws used as standoffs. The night-light shown in Fig. P4-2 uses a 4 inch by 1 1/2 inch by 1/16 inch copper sheet bent at right angles on inch from one end as a heat sink. Five holes have been drilled into this heat sink; a 1/4-inch diameter hole on the bent-up end for the triac and four number 8 clearance holes for the nylon screws. Matching holes for the nylon screws have been cut in the top of the minibox. These four number 8 holes also hold the two four-point terminal strips. After the triac has been attached to the heat sink, the nylon screws are inserted through the box. Several fiber or nylon washers are placed over the screws to keep the heat sink isolated from the box. The heat sink is placed on the screws. Four number 8 nuts hold the entire assembly secure. The nuts do not have to be nylon.

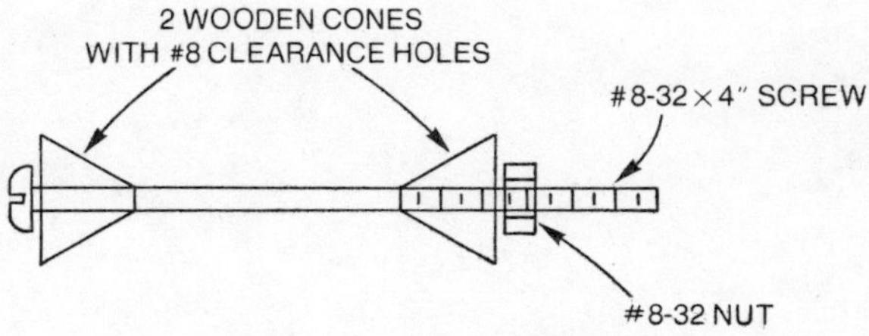

Fig. P4-3. A simple coil-winding jig suitable for winding the reed switch coil.

The coil for the reed switch is available commerically or you can make your own. The coil form is supplied with the reed switch. Ten thousand turns can be accomplished rather easily with the aid of a small winding jig and a drill. Figure P4-3 shows a simple coil-winding jig which will hold the coil form. The coil wire is very fine and care must be taken to avoid kinks or breaks. When the coil is wound, wrap the coil in electrical tape to hold the turns in place.

When you have completed the assembly and checked for errors, plug a lamp into the load receptacle and the line cord into a convenient AC outlet. With daylight falling on the photocell the lamp should be out. With no light or little light on the photocell the lamp should come on. With this unit placed in a window, the lamp will come on when it gets dark and turn off at dawn.

(This project courtesy of General Electric, Semiconductor Products Department.)

5

Nighttime Warning Flasher

With battery-operated flashers it is important to keep the drain on the battery down as low as possible. One easy way to do this is to turn the flasher off during the daylight hours. This circuit is small and convenient to use around buoys, piers, docks, and towers far from 120 VAC power. It is also useful as a highway marker for parked vehicles.

The circuit is shown in Fig. P5-1. With daylight on the phototcell, its resistance is very low. This allows only a low voltage to the gate of the SCR, too low in fact to trigger the SCR. When the photocell is dark its resistance is much higher and, therefore, there is sufficient current through R2 to trigger the SCR. When the SCR triggers at about five volts, it turns on the lamp. When the lamp's light illuminates the photocell, the photocell's resistance drops again to a low value. This allows the capacitor to charge to five volts in the polarity shown in the figure. The lamp used in this circuit contains a bimetal element so that when heated, the lamp opens and turns off the current through the SCR. As the bimetal cools, it closes the internal contacts in the lamp. The photocell with no light on it returns to its high resistance state. This allows the capacitor to discharge and charge to one volt in the opposite polarity. This capacitor voltage retriggers the SCR. The lamp "off" time is

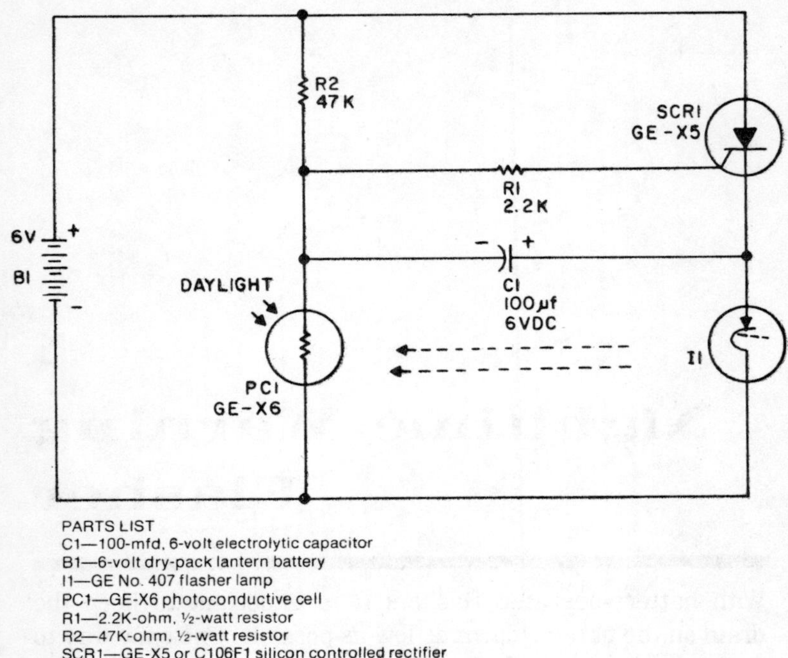

Fig. P5-1. Automatic nighttime warning flasher schematic and parts list. (Courtesy of General Electric Semiconductor Products Dept.)

about five seconds. This can be reduced by reducing the capacitor, but at the expense of shorter battery life.

The project should be built within a water-tight box if it is to be used near the water. If it will be used near seawater, special enclosures and construction techniques are required, as most who live by the ocean are aware. The lamp and photocell should be mounted on the same surface of the enclosure so that the photocell can receive the lamp's light as explained. The balance of the components can be easily mounted on either perforated board or terminal strips. Polarity must be observed for the capacitor and SCR.

6
Logic Probe

With many of us building projects with TTL integrated circuits, it is very handy to have a convenient indicator to tell us the logic state of the different gates within a circuit. This logic probe can be built within an old penlite flashlight and is used in a manner similar to the neon test lights for 120 volts AC.

The circuit, as shown in Fig. P6-1, is a simple transistor inverter/amplifier stage. When voltage in excess of two volts is applied between the ground lead and the probe, the darlington transistor turns on and energizes the LED. Therefore a high logic state is indicated by the LED being on, a low logic state by the LED being off.

The project can be built in any long tubular container such as a cigar tube, pill vial, or a penlight. I chose the penlight since it is the correct diameter for the batteries and contains the spring contact along with a mounting place for the LED. A penlight in which the end opposite the lamp opens for battery replacement is best but not necessary. Mine was an Eveready which had been taking up space in the workbench drawer for ten years.

To build the unit, start by drilling one or two holes in the side of the penlight about one-third the distance down from the

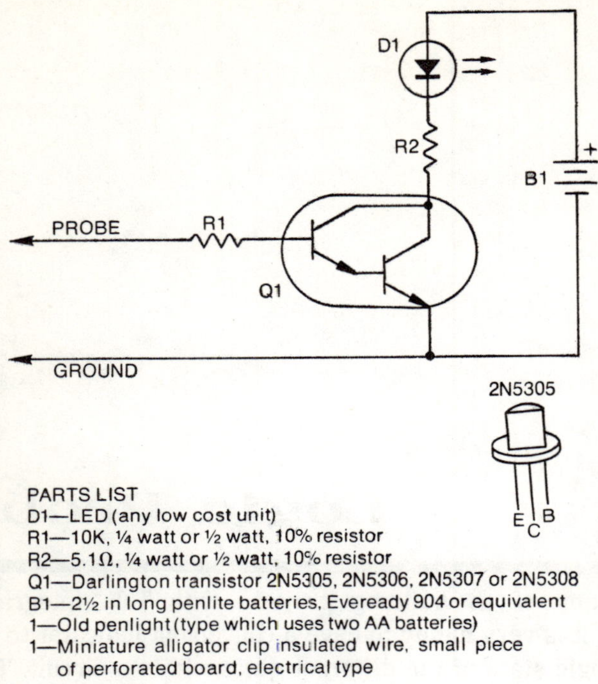

Fig. P6-1. IC logic probe schematic and parts list.

lens. The holes should be just large enough for the insulated ground and probe wires to pass through. Remove the lamp from the penlight, modify to make a hole large enough for the LED to slip through where the lamp would normally be, and reassemble. Cut a piece of perforated board the length of a penlight cell and wide enough to fit snugly within the battery compartment. The LED is mounted on one end of this board so that it will protrude through the end of the penlight where the bulb was. The resistor R2 and transistor Q1 are then wired to the board and LED. A heavy wire from the transistor's emitter lead should be layed parallel to the length of the board with a slight bow. This wire will act as a spring contact to the case of the penlight, which is the battery's negative terminal. A 12-inch length of flexible insulated wire is attached to the spring wire. This wire will become the ground wire for the lead of the transistor. The second wire will later be attached to the probe resistor.

The final work on the board is to run a wire the length of the board from the LED's positive lead to the opposite end. This wire is wrapped over the opposite end at the midpoint of the end and secured. Solder should be liberally applied over the end to build up a contact to the battery's positive terminal as shown in Fig. P6-2. No part of the circuit board's electrical circuit should touch the case of the penlight when inserted except for the ground contact. Some electrical or Mylar tape is a good insulator.

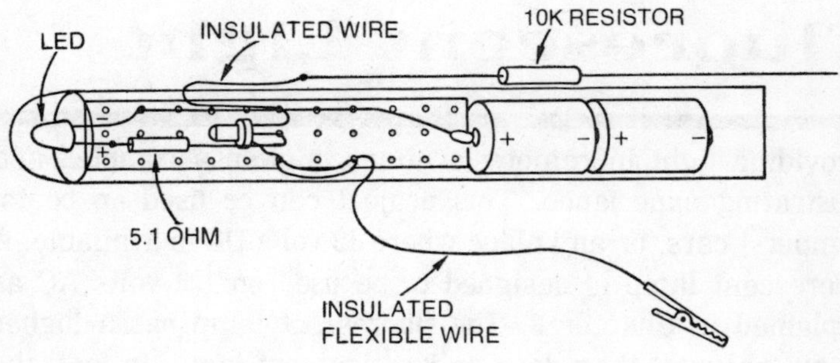

Fig. P6-2. Typical parts layout in a penlite case. The text has details on the construction.

The probe wire and the ground wire are then passed through the holes in the side of the flashlight and the board is slid into place. The penlight case should then be wrapped in electrical or Mylar tape. The ground wire should be placed under one turn of tape to prevent it from being pulled out of the assembly.

The probe resistor should be taped to the side of the case so that one lead extends about one inch past the battery end. The probe wire is then soldered to the other end of the resistor. A final few turns of electrical tape should be used to cover any bare conductors from the probe wire. Attachment of the alligator clip to the ground wire completes the logic probe.

When the batteries have been inserted, attach the ground lead to the negative of the logic power supply and probe away.

7

Battery Operated Fluorescent Light

Providing light in remote locations is often a complex and frustrating experience. This project can be used on boats, campers, cars, or any place where 12 volts DC is available. A fluorescent lamp is designed to be used on 120 volts AC as explained in Chapter 3. The fluorescent lamp has a higher output-per-watt than does an incandescent lamp. In fact, the 8W fluorescent lamp used in this project is equivalent in light output to a 30-watt incandescent bulb. This means that a standard 60 ampere-hour car battery could run the 2-lamp circuit for over 30 hours, but would run the equivalent light from two 30-watt incandescent bulbs for only one hour.

The circuit is a single transistor inverter. The basic circuit was developed by Jack Campbell at General Electric's Nela Park many years ago and has been used in several commercially available battery-operated fluorescent lamps. The power transistor pulses the input voltage across winding N3 of the transformer. This transformer is an auto-transformer. At point A the voltage pulses are about 280V. The capacitors C4 and C5 serve to block out the DC component of the point A voltage waveform. This presents only AC to the lamps. Capacitors C4 and C5 are the actual ballasts for the lamps. Capacitor C3 acts to regulate the output voltage. Capacitor C2 and R2 serve only to start the inverter when power is applied, while C1 and the number 47 lamp serve to regulate base current to the power transistor.

A 12 × 2½ × 2⅛-inch minibox is slightly larger than the F8T5 lamps which are used, and was chosen as the mounting base for the project. Any small fluorescent fixture which uses these bulbs can be used and would save the mounting of the lamp sockets and the assembly of a reflector. With the minibox, a reflector was made out of an 11 × 5-inch piece of bright aluminum. The sides were bent up about 30° and attached to the top of the minibox. The lamp sockets were then attached (two on each end for a two-lamp circuit or one on each end for a single lamp). The addition of the on/off switch completes the nuts and bolts work.

The balance of the components can be mounted on Vectorboard as shown in Fig. P7-2. The power transistor should be mounted to the minibox by the following procedure. First, an inch long section of electrical tape is attached to the inside of the minibox near the Vectorboard. A small amount of epoxy glue is then applied over the tape, and the power transistor's flat back is centered onto the tape and pressed firmly. After the glue has set, a quick check with an ohmmeter (from the tab to the minibox) should show high impedance. This mounting method provides an isolated mounting as well as sufficient heat transfer for the transistor.

While awaiting the epoxy on the transistor to cure, you can wind the transformer. Start with 35 feet of No. 32 AWG magnet wire and place two hundred turns of wire on the core in two layers of 100 turns each. After each layer, cover the layer with a single layer of electrical or Mylar tape. Mark the start and finish ends of the winding for N1. Now with a couple more feet of No. 32 AWG magnet wire, wind six turns for N2. Wind in the same direction as before and again mark start and finish of N2. The N2 winding also deserves a bit of electrical tape to hold it in place. The last winding, N3, is made of eight turns of No. 20 AWG magnet wire, wound again in the same direction and with the start and finish marked. Now for the connecting of the windings. Figure P7-1 shows the transformer with each winding (N1, N2, and N3) marked for start and finish. Solder together N1 start and N2 finish, then repeat with N2 start and

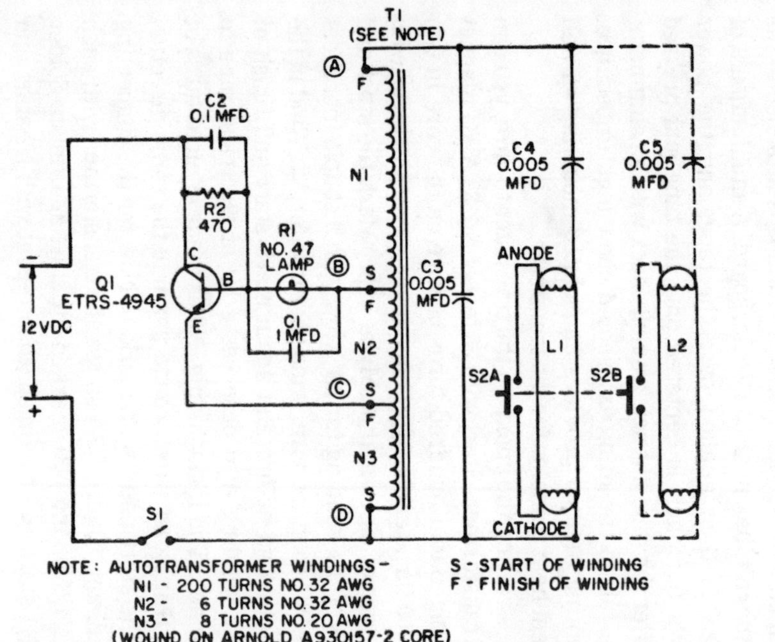

Parts List

C1 — 1-mfd, 50-V paper capacitor
C2 — 0.1-mfd, 50-V paper capacitor
C3, C4, C5 — 0.005-mfd, 1000-V disc-ceramic capacitors (delete C5 for single lamp)
*Q1 — ETRS-4945 transistor**
R1 — GE No. 47 bulb
R2 — 470-ohm, 1/2-watt resistor
L1, L2 — GE F8T5-CW fluorescent lamps
S1 — SPST toggle switch
S2 — DPST momentary push button switch (if only one lamp is needed, a SPST push button is used)

T1 — Autotransformer — core available from GE, ETRS-4891. See text for winding details.*
Minibox — 12" x 2-1/2" x 2-1/4" (Bud CU-2114-A), or equivalent
Pin Sockets — GE ALF141-33, or equivalent

**Available from General Electric Co., Dept. B, 3800 N. Milwaukee Ave., Chicago, Ill. 60641*

Fig. P7-1. Parts list and schematic diagram for the battery-operated fluorescent lamp. (Courtesy of General Electric.)

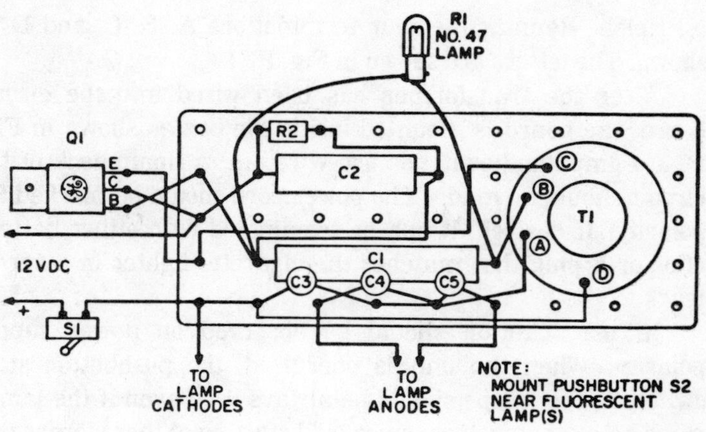

Fig. P7-2. The majority of the components can be assembled on a perforated board as shown above. (Courtesy of General Electric.)

GE F8T5-CW
FLUORESCENT
LIGHT

BATTERY CABLE
WITH CLIPS

LAMP
R1

MINIBOX
12" x 2-1/2" x 2-1/4"

T1

CIRCUIT
BOARD

Q1

S1 AND S2

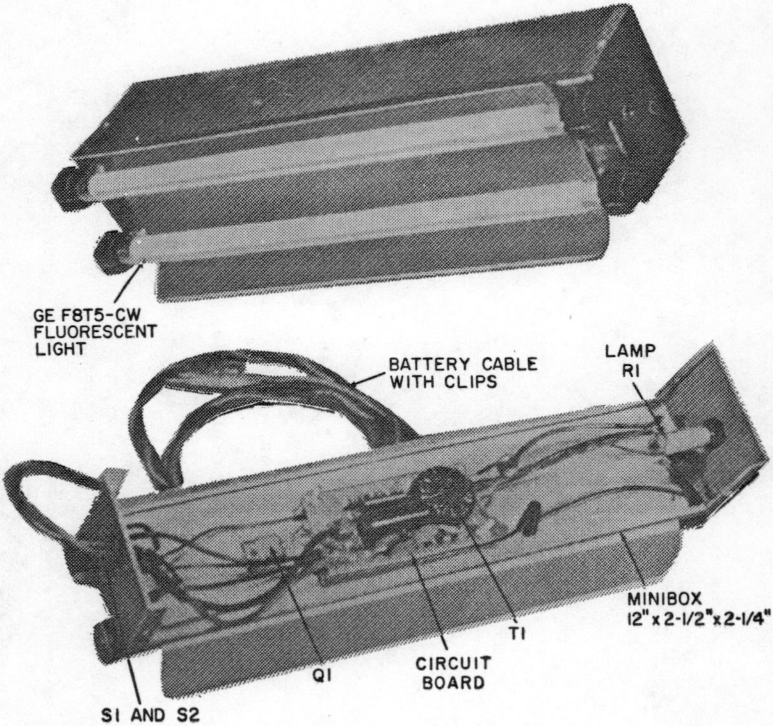

Fig. P7-3. Top and bottom view of a completed battery-operated fluorescent lamp. (Courtesy of General Electric.)

N3 finish. Remark the four terminations A, B, C, and D as shown. The letters are shown in Fig. P7-1.

After the transformer has been wired into the circuit board, the board is mounted inside the box as shown in Fig. P7-3. Lamps and switches are wired in. A final check of the circuit should be made. The power cord should be marked for polarity. If desired, it can be terminated with either battery clips or a plug that matches the cigarette lighter in a car or truck.

In use, caution should be observed on power supply polarity. When the unit is energized, the pushbutton start switch on the lamps should be always used even if the lamps come on by themselves, since cold starting of these lamps will severely limit their life.

(This project courtesy of General Electric Company.)

8

Lazy-Man's Switch

After watching the late-late show through the toes, I sure hate to get up and turn off the television. Unlike most, I did something about it. Now all I do is pick up the flashlight on the night stand and flash it on the "Lazy-Man's Switch," which sits on top of the television. This switch is set up so that it can turn on or off whatever load is plugged into the outlet on the back of the box, either by flashlight at a distance or by the pushbutton switches on top of the unit.

The circuit, as shown in Fig. P8-1, uses a triac to turn the load on and off. The balance of the circuit is the sensing and trigger circuit for the triac. SCR1 and SCR2 are connected as an SCR flip-flop. When SCR1 is on, gate current is supplied to the triac through R8. When SCR2 turns on, SCR1 is turned off by the stored charge on C3. C3 should be a film or foil type of capacitor, not an electrolytic capacitor. Either SCR can be triggered by closing the pushbutton switch connected to its gate. These switches provide a means of turning the load on and off at the "Lazy-Man's Switch" without the flashlight. (If you are a truly lazy man, you don't want to walk across the room for a flashlight.)

When the "Lazy-Man's Switch" is triggered with light, the unijunction and light-sensitive transistor circuit (Q1, Q2) does

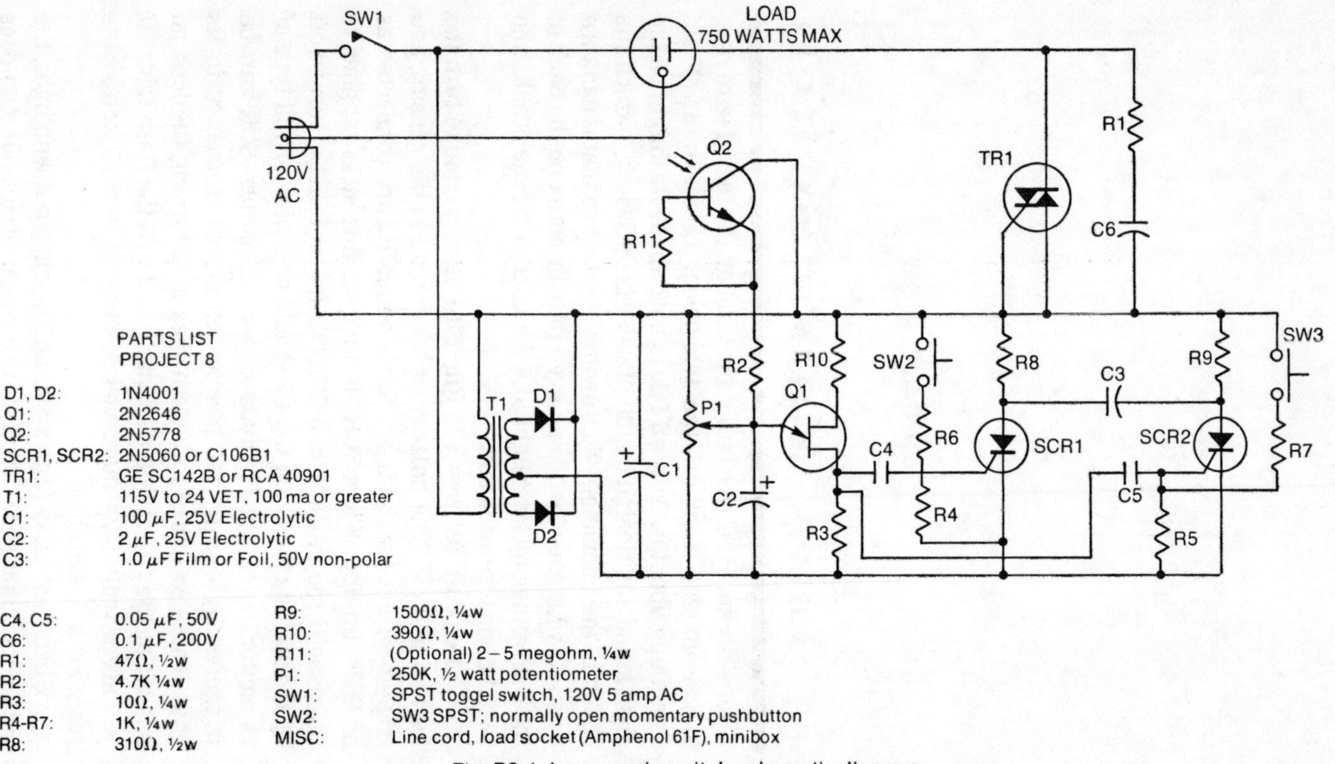

PARTS LIST
PROJECT 8

D1, D2:	1N4001
Q1:	2N2646
Q2:	2N5778
SCR1, SCR2:	2N5060 or C106B1
TR1:	GE SC142B or RCA 40901
T1:	115V to 24 VET, 100 ma or greater
C1:	100 μF, 25V Electrolytic
C2:	2 μF, 25V Electrolytic
C3:	1.0 μF Film or Foil, 50V non-polar
C4, C5:	0.05 μF, 50V
C6:	0.1 μF, 200V
R1:	47Ω, ½w
R2:	4.7K ¼w
R3:	10Ω, ¼w
R4-R7:	1K, ¼w
R8:	310Ω, ½w
R9:	1500Ω, ¼w
R10:	390Ω, ¼w
R11:	(Optional) 2 – 5 megohm, ¼w
P1:	250K, ½ watt potentiometer
SW1:	SPST toggel switch, 120V 5 amp AC
SW2:	SW3 SPST; normally open momentary pushbutton
MISC:	Line cord, load socket (Amphenol 61F), minibox

Fig. P8-1. Lazy man's switch schematic diagram.

the job of the pushbutton switches. If we assume that SCR2 is on, then the load is off. When light is shone on Q2, it turns on the supplies additional charge to capacitor C2. Unijunction Q1 then triggers, giving a trigger pulse to each SCR. Since SCR1 is off, it triggers. This both energizes the load and turns SCR2 off. The next pulse from the unijunction transistor reverses the process, turning SCR2 on and SCR1 off, and thereby turning the triac and load off. The unijunction is set so that it will not oscillate, by adjusting the 250K potentiometer so that the capacitor voltage, without light, is just below the threshold of the unijunction. The current that is supplied by the phototransistor is then sufficient to trigger the unijunction.

Power for the trigger circuit is derived from the transformer–rectifier circuit. In this circuit, a full-wave centertap configuration has been used. The capacitor C1 serves as the power-supply filter.

The construction of this project should start with the assembly of the circuit board. Everything except the load socket, switches, triac, phototransistor, and transformer can be mounted directly on a piece of perforated board. Once the board is completed, the minibox can be prepared by mounting the switches, load socket, transformer and triac on the inside of the box in convenient locations. Holes should also be included for the line cord and phototransistor. The triacs listed contain internal isolation, so mounting tabs are electrically isolated from the three leads. If a different type is used, caution should be observed when mounting the triac. It is important to heat-sink the triac by either attaching it to the minibox or a heat sink. It is also of the utmost importance to isolate any live electrical components from the minibox.

Mount the phototransistor by drilling a hole in the front of the minibox. Insert a ¼-inch ID rubber grommet in the hole and insert the phototransistor in the grommet with the rounded side of the transistor centered within grommet. The transistor can be held in place with epoxy glue. (Take care not to get the epoxy on the phototransistor leads as this will cause problems when soldering the leads.) The phototransistor mounted this way is protected from ambient light and is therefore less susceptible to being triggered by room light.

Once the circuit board is installed in the box and the other components are wired into the circuit, an assembly check should be made to insure that the proper connections have been made and that no live electrical parts are in contact with the minibox. The unit can be checked by plugging a lamp into the load socket. Either the *off* or *on* pushbutton switch will set the unit. Turn the potentiometer slowly. At a third of the distance from one end the lamp should flash at a rapid rate. Keep turning the potentiometer until the flashing slows down and eventually stops. At this stopped point the circuit is most sensitive, and a flashlight shone on the phototransistor for a brief moment should cause the lamp to turn on or off. The flashlight should be effective up to a distance of fifteen feet.

If high ambient-light levels create problems using this unit, the phototransistor can be desensitized with a resistor across its base-emitter junction. The value of this resistor should be two to five megohms.

9

Touch Switch

To many of us the electromechanical switch on our lamps and appliances is rather ugly. In addition, manufacturers put them in the worst places in the world. You need to be a carnival rubber-man to get at them. With this touch switch built into a lamp, you need only brush your hand across the contacts and the lamp turns on. Brush again and the lamp is off. The circuit shown here is "boxed," but it can easily fit in the base of most table and floor lamps or even in a wall junction box.

The circuit is shown in Fig. P9-1 and uses an SCR to drive the lamp. Add a relay for lamp loads over 150W if you mind the lamp being dimmer due to the half-waving through the SCR. With the relay, lamps and other loads up to 500W can be controlled.

Upon plugging the circuit in, lamp L1 and load J1 are off. When the touch switch contacts are bridged by a low impedance, such as a finger, base current is provided for phototransistor Q1. Q1 turns on and causes Q2 to turn off. This in turn causes the SCR to be triggered. Lamp L1 comes on and illuminates the phototransistor, latching the phototransistor on even after the removal of the impedance across the touch plate.

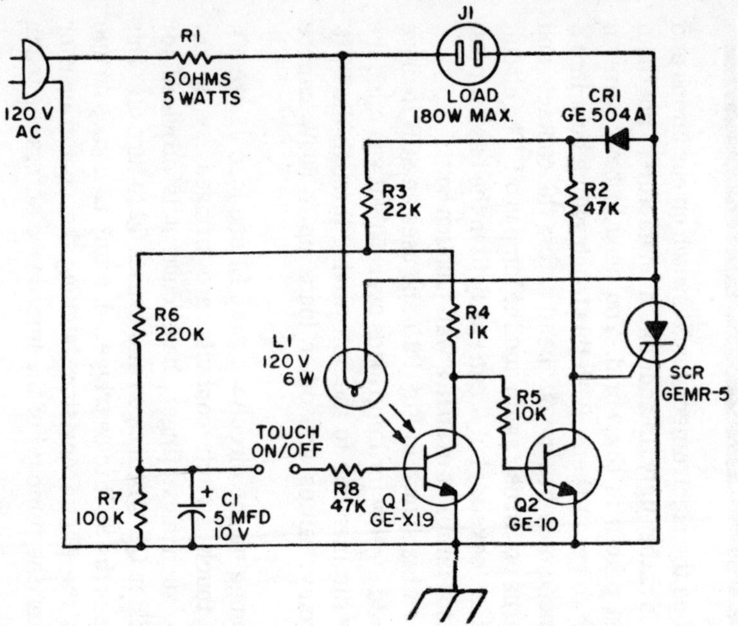

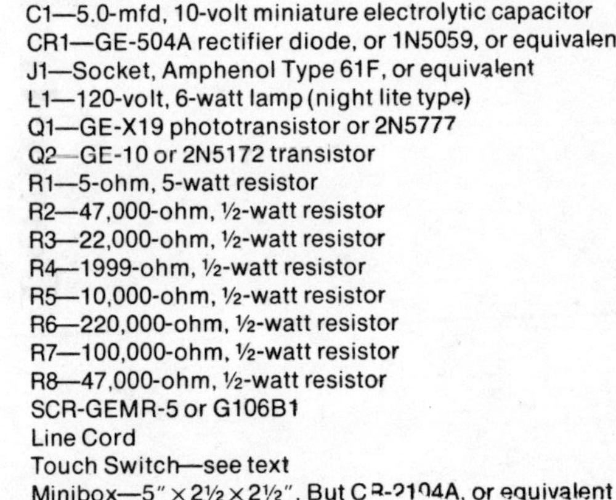

PARTS LIST
C1—5.0-mfd, 10-volt miniature electrolytic capacitor
CR1—GE-504A rectifier diode, or 1N5059, or equivalent
J1—Socket, Amphenol Type 61F, or equivalent
L1—120-volt, 6-watt lamp (night lite type)
Q1—GE-X19 phototransistor or 2N5777
Q2—GE-10 or 2N5172 transistor
R1—5-ohm, 5-watt resistor
R2—47,000-ohm, ½-watt resistor
R3—22,000-ohm, ½-watt resistor
R4—1999-ohm, ½-watt resistor
R5—10,000-ohm, ½-watt resistor
R6—220,000-ohm, ½-watt resistor
R7—100,000-ohm, ½-watt resistor
R8—47,000-ohm, ½-watt resistor
SCR-GEMR-5 or G106B1
Line Cord
Touch Switch—see text
Minibox—5″ × 2½ × 2½″, But CR-2104A, or equivalent

Fig. P9-1. Touch switch schematic diagram and parts list. (Courtesy of General Electric Semiconductor Products Dept.

When the SCR is on, there is no voltage supply to keep capacitor C1 charged and it quickly discharges. If the touch plate is bridged again the base of Q1 is essentially grounded, causing a large decrease in the light gain of the phototransistor. With the values and layout shown, Q1 comes out of saturation and allows Q2 to saturate, turning off the SCR.

Notice that if the touch plate is permanently shorted, the lamp will flash on and off at several cycles per second.

The assembly of the circuit can be done as shown in Fig. P9-2. Before inserting Q1 into the circuit board, note that the package is transparent and that one side is rounded. If you look closely into the rounded side you will see a small silver-colored rectangle. This is the actual phototransistor; the rest of the device is just packaging to make it easier for you to handle. It is important that Q1 be inserted into the circuit board with the transistor chip squarely facing the 6W lamp.

The minibox needs to have four holes drilled into it. One large one for the socket, one small one for the board mounting

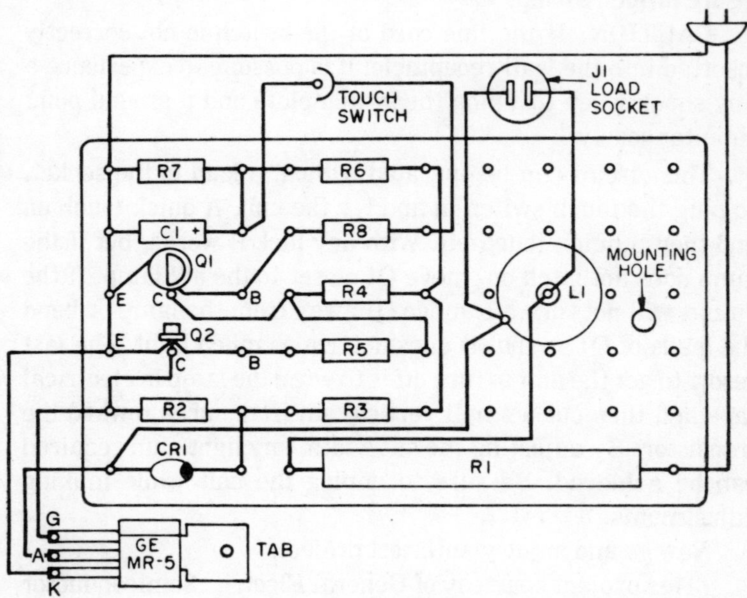

Fig. P9-2. Perforated board layout for the touch switch. (Courtesy of General Electric Semiconductor Products Dept.)

screw, another for the touch switch leads, and the fourth for the line cord. It is best to use a rubber grommet to keep the line cord from being cut by the metal box.

The touch plate can be made by etching two interwoven but isolated fingered runs on a printed circuit board. If you do not have etching facilities, lay down an even one-inch square of electrical tape on the top of the minibox. Coat the tape with a uniform coat of epoxy. Lay two pieces of 22 AWG bus wire formed into an "S" shaped pattern into the epoxy so that the wires are parallel but not touching. After the epoxy has set, carefully sand the epoxy and wire until the bus wire is exposed for its full length. Two small flexible wires should then be attached to the bus wire and fed to the inside of the box.

Repeat the application of a one-inch square of electrical tape coated with epoxy on the inside of the minibox. The heat sink tab of the SCR should be pressed firmly into the epoxy and the epoxy allowed to cure.

After attaching the SCR, touch plate, socket, and line cord to the perforated board mounted inside the box, check to insure proper wiring.

CAUTION: If the line cord of the switch is not correctly inserted into the wall receptacle, it is possible to experience a tiny shock when touching the switch plate and a ground point simultaneously.

This circuit can be operated without a load in the socket, so plug the touch switch in and try the unit. A quick touch on and then a quick touch off. With any luck it works, but if the lamp does not latch on, move Q1 closer to the 6W lamp. If the circuit will not turn off, move Q1 away from the lamp or bend the leads of Q1 so that it does not see as much light. The last resort to get the unit to turn off is to wrap the lamp in electrical tape and then cut a small vertical slit in the area toward the transistor. By adjusting the slit width, any light gain required can be achieved. Be sure to unplug the unit while making adjustments.

Now go and enjoy your latest project.

(This project courtesy of General Electric, Semiconductor Products Dept.)

10

High Performance Color Organ

One of the all-time favorite electronics projects is the color organ. It has been called many names, but basically it is a circuit that turns on and off in response to music; different frequencies cause different sets of lights to turn on and off. There have been many of these circuits published, but this circuit has very good performance for the cost. The system as shown will drive up to 1800W of lamps; 600W on each of the three channels.

The color organ is to be connected to the output of the power amplifier. The 250-ohm potentiometer, P4, serves as the master sensitivity control for the entire system. It divides the amplifier audio output to achieve the correct voltage input levels for different sound levels. The audio transformer serves to electrically isolate the amplifier ground from the color-organ ground and to transform the voltage and impedance levels for the filters. The three filter networks divide the audio into the three frequency bands. The filters are straightforward circuits with only one trick used. Choke L4 in the midrange filter is a winding of an audio output transformer used as a choke. It has the right characteristics, works well, and is reasonably inexpensive, so why not?

The filters drive three identical circuits. The output of the filter drives the base of a transistor, which charges the 0.47 μF

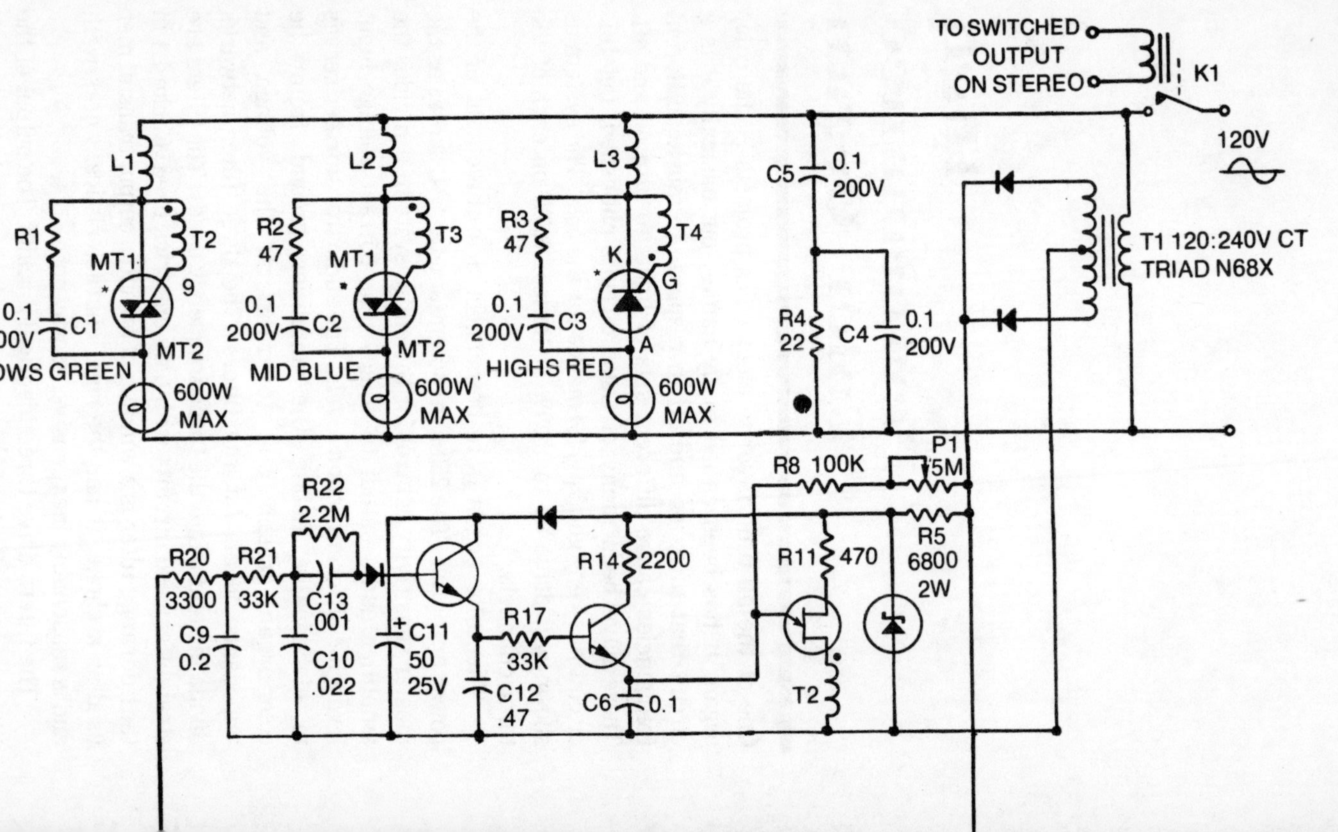
TO SWITCHED
OUTPUT
ON STEREO
K1
120V
T1 120:240V CT
TRIAD N68X
L1
L2
L3
R1
T2
MT1
9
0.1
200V
C1
MT2
LOWS GREEN
600W
MAX
R2
47
T3
MT1
0.1
200V
C2
MT2
MID BLUE
600W
MAX
R3
47
T4
K
G
0.1
200V
C3
A
HIGHS RED
600W
MAX
C5
0.1
200V
R4
22
C4
0.1
200V
P1
5M
R8 100K
R5
6800
2W
R22
2.2M
R20
3300
R21
33K
C13
.001
C9
0.2
C10
.022
C11
50
25V
R17
33K
C12
.47
R14
2200
C6
0.1
R11
470
T2

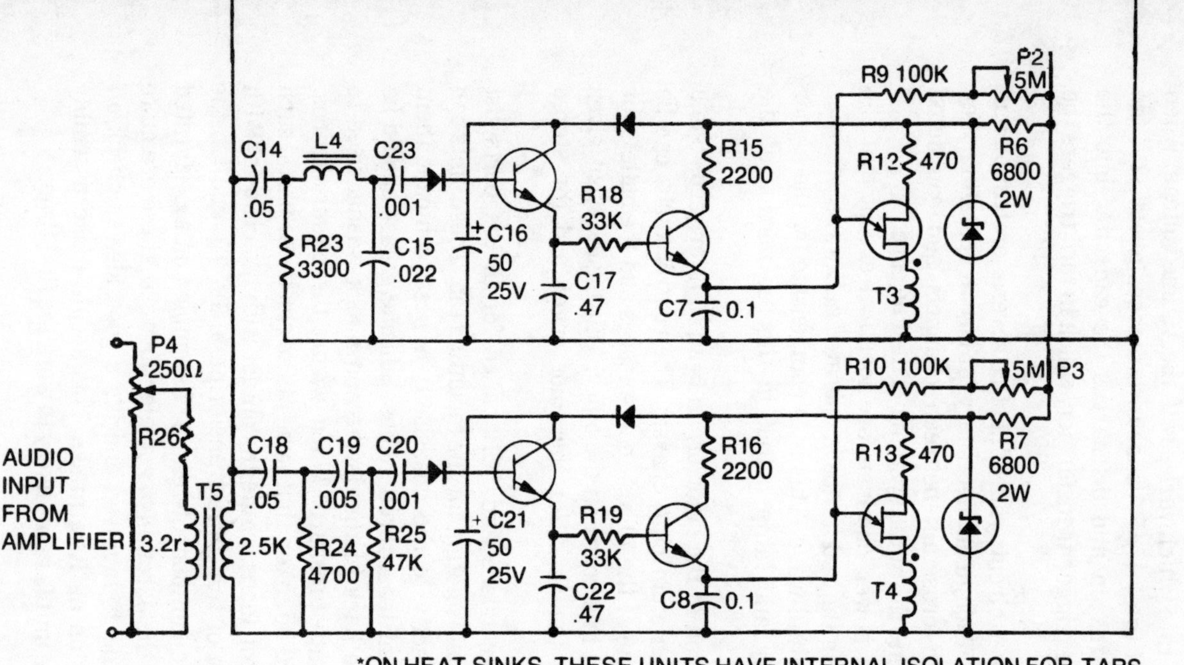

*ON HEAT SINKS, THESE UNITS HAVE INTERNAL ISOLATION FOR TABS

Fig. P10-1. Schematic of the high performance color organ.

capacitor to the peak voltage of the signal that appears on the base of the transistor. The second transistor acts like the first and charges the 0.1 μF capacitor of the unijunction-transistor trigger circuit to a voltage that is directly proportional to the voltage of a given frequency band within the audio input. The five-megohm potentiometer adds additional voltage to the capacitor at a rate which varies with the AC line voltage. When the voltage on the 0.1 μF capacitors reaches about 12V, the unijunction turns on and discharges the capacitor into the primary of the trigger transformer, which in turn triggers the triac or SCR.

In the power circuit you will notice several protection and filter circuits. Though not necessary for circuit operation, they minimize both crosstalk between channels and conducted interference into the audio system via the line cord.

The three power channels are identical except that the channel driving the red lamps is operated half-wave, whereas the others are full-wave. This is to compensate for the higher eye sensitivity and lamp output in the red portion of the spectrum.

Construction of the color organ is straightforward. Use a 10 × 6 × 3-inch aluminum chassis. The control knobs can be mounted on one of the 10 × 3-inch surfaces and the outlets for the lamps and line cords on the other. This basic layout gives you room on the shelf for the color organ and the stereo components.

The power triacs and SCR called for in the parts list contain internal isolation for the mounting tab. These units can be bolted directly to the chassis. It is best to mount them separated by at least three inches. Silicone grease should be applied to the back of the device and a lock washer used to insure adequate pressure. When wiring the power devices, caution should be taken that none of the electrical parts, such as the thyristor leads, come in contact with the chassis. With the exception of the three transformers, T1, T5, and L4, the balance of the components can be assembled on a perforated board. While the assembly is straightforward, there are quite a few components involved and it's a good idea to check off each connection as it is made. This should make assembly easier and the unit is more likely to work the first time.

PARTS LIST
CAPACITORS
C1-C8: 0.1 μF, 200V paper or film capacitors
C9: 0.2 μF, 50V
C10, C15: 0.022 μF 50V
C11, C16, C21: 50 μF, 25V electrolytic
C12, C17, C22: 0.47 μF, 25V
C13, C20, C23: 0.001 μF, 25V
C14, C18: 0.05 μF, 25V
C19: 0.005 μF, 25V

RESISTORS (all ± 10%, ¼W unless noted)
R1-R3: 47Ω, ½W
R4: 22Ω, ½watt
R5-R7: 6800Ω, 2W
R8-R10: 100K, ½W
R11-R13: 470Ω
R14-R16: 2200Ω
R17-R19, R21: 33K
R20, R23: 3300Ω
R22: 2.2 meg.
R24: 4700Ω
R25: 47K
R26: 6.8Ω, 5W

POTENTIOMETERS
P1-P3: 5 meg, 2W
P4: 250Ω, 2W

INDUCTORS
L1-L3: 100μH, 5A
L4: Triad S-12X or Stancor A-3332 secondary used as choke

TRANSFORMERS
T1: 120V: 240 VCT, Triad N68X or equivalent
T2-T4: SCR 1:1 trigger transoformer, Spraque 11Z12 or equivalent
T5: Triad S-12X or Stancor A-3332

SEMICONDUCTORS
(3) Unijunctions: 2N2646 or GE-X10
(6) Transistors: 2N5172, GE-10, or 2N222
(8) Rectifiers: 1N4004 or 1N5059 (200V, 1A)
(3) Zeners: 1W, 16V
(2) Traics: GE SC142B or RCA 40901*
(1) SCR: GE C123B*

*mount on Heatsink

Misc.: Lamps and lamp holders or sockets
line cord
aluminum chassis 6×10×3 with bottom cover
perforated board
heatsink(s)
K1: 120V coil, 15A relay (optional)
jack for the audio input

P-10-1. Legend.

After the color organ is assembled, checked, and rechecked, connect the unit first to the audio source then plug it in to the 120 VAC. If you are using the remote switch, make this connection also. With the system under power, and any lamp load, set the 5 meg potentiometers so that the lamps are just off with no audio input. Each of these potentiometers should give a response like a lamp dimmer. If not, then there is a wiring error in the power or relaxation-oscillator sections of that channel. Once these sections are set, audio should be applied to the audio input. If you have an audio oscillator you can check each channel in turn by varying the input frequency. If you do not have an oscillator, a music source which contains a wide range of frequencies should be used. It will be necessary to observe the lamps in response to the music to determine if all the channels are functioning properly. The master control allows you to increase the input to the color organ when the audio system is playing at low volume, or to decrease the input when it is too high for the color organ to respond correctly.

The brilliance of the display also depends on the lamps used. The smaller the lamp, the more brilliant the display. A combination of lamps of different sizes can cause additional interesting effects. So the choice of lamps is open. The author has used a matrix consisting of 9 strings of 35 miniature lamps each. These are standard miniature series-wired Christmas lights. A three-foot square frame was constructed from 1 × 4-inch pine. Inside the frame was placed four pieces of 1 × 1-inch strips so that there was ⅛-inch to the back of the frame (see A in detail of Fig. P10-2). A spare piece of wood paneling was cut to provide two squares which would fit snugly into the frame on each side of the strips. The panel for position C (detail B) will form the back, while the panel for position B will be the mounting panel for the 315 lamps. Once panel B is cut to size, drill snug-fit holes for each lamp. The panel is then mounted into the frame and the number one strips tacked in to hold the panel in place.

At this point, the fun begins. Take each string of lights and, after insuring that it works and all the lamps are good, remove all the lamps and separate them according to color. Now if you

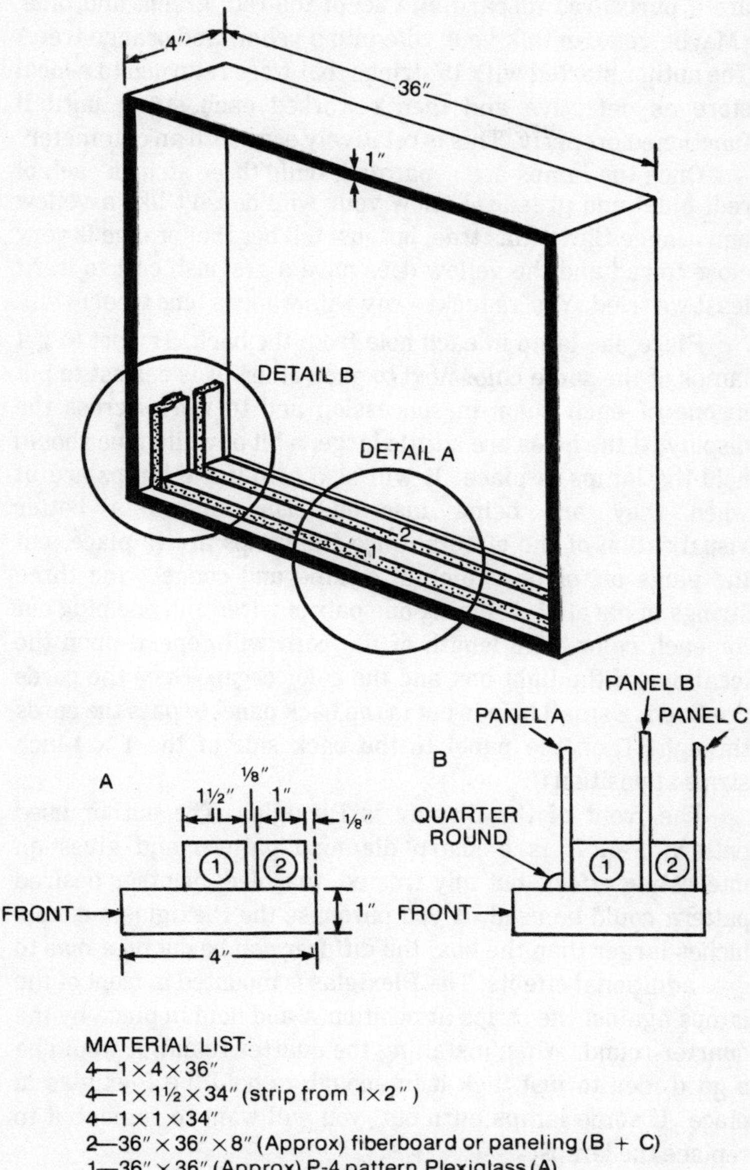

Fig. P10-2. Construction of the light box display for your color organ. The detail shows the positioning of the panel supports.

are a purist you discard all except the red, green, and blue. (Maybe you can talk your wife into a yellow and orange tree.) The author started with 15 strings that were returned to a local store as defective and then reworked each string until it functioned properly. This is relatively easy with an ohmmeter.

Once the lamps are separated, build three strings each of red, blue, and green. (I know your wife doesn't like a yellow and orange Christmas tree, but just tell her that orange is very close to red and the yellow does have a greenish cast to it. At least we tried. You're lucky—my wife wants a blue tree!)

Place one lamp in each hole from the back. Try not to get lamps of the same color next to each other. It is easiest to put in one of each color in succession and to work across the display. If the holes are a little large, a bit of white glue should hold the lamps in place. It will also help if the lamps are lit when they are being inserted. This allows a better visualization of the effects. Once the lamps are in place, cut the plugs off of one color at a time and connect the three strings in parallel, bringing one pair of wires and one plug out for each color. The length of the cord will depend upon the locations of the light box and the color organ. Once the cords are fixed, a small hole is cut in the back panel to pass the cords through. Tack the panel to the back side of the 1 × 1-inch strips at position C.

The front of the display is Plexiglas. The author used pattern P-4. It is a sharp diamond pattern and gives an interesting effect, but any frosted, marbled, or other desired pattern could be used. If you purchase the Plexiglas a couple inches larger than the box, the diffuser can be cut on a bias to give additional effects. The Plexiglas is mounted in front of the lamps against the strips at position A and held in place by the quarter-round. When installing the quarter-round, it would be a good idea to just tack it in enough to hold the Plexiglas in place. If some lamps burn out, you will want to remove it to replace the lamps.

The light box is now functional; this is where the cabinet maker must take over to finish the surfaces.

Other effects are possible through various lighting systems. For example, the low-frequency channel could be connected to a fluorescent dimming ballast. The fluorescent lamp does not have the long response time of the incandescent, so that the low-frequency channel will become more of a stroboscopic effect.

As an added feature, relay K1, when slowed to the remote turn-off switch on the stereo, will turn off the color organ when the stereo master switch is turned off.

11
Strobe Light

Stop-motion photography and psychedelic lighting call for strobed lighting effects. The circuit shown in Fig. P11-1 is capable of strobe rates from about one to five pulses per second, and with minor modifications it can be extended to lower or higher flash rates. The intensity is kept lower than the slave flash of Project 3 to extend the life of the flashtube, but it should be bright enough for most applications.

The circuit consists of a voltage-doubler power supply and a relaxation-oscillator trigger circuit. The battery supply used in Project 3 can be substituted if desired. Since similar components are used in the two projects, the projects can be combined with a multiple-pole switch to give a dual-function circuit. (This is not presented in this book, but should not be beyond the skills of the more advanced hobbyists.)

The line voltage drives the voltage-doubler circuit directly. An isolation transformer (such as a Stancor P-6411) can be added if desired between the switch and the diode doubler. C3 serves as the energy-storage capacitor for the xenon lamp L1, while C4 provides the energy for triggering the lamp. The lamp is triggered when the SCR is turned on and discharges the capacitor C4 through the trigger transformer. The circuit is designed so that the SCR is triggered at regular intervals by the discharging of C5 through the neon lamp. The

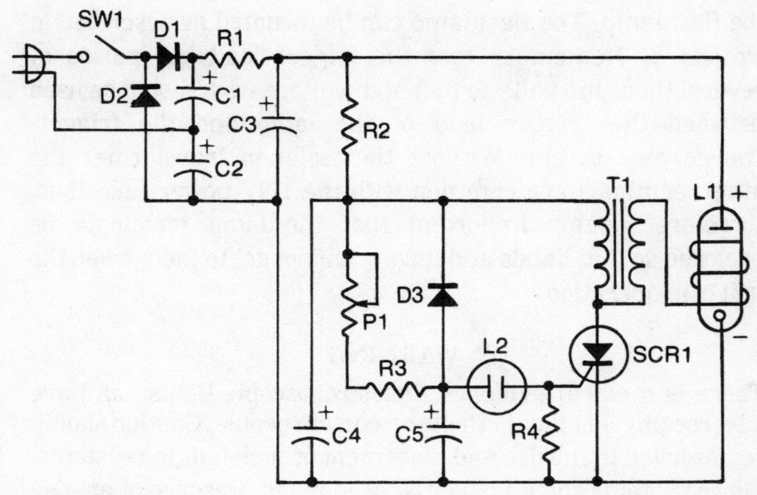

PARTS LIST
D1-D3: 1N5060 or 1N4004
C1-C2: 5 μF, 200 VDC, electrolytic
C3: 25 μF, 400 VDC,, electrolytic
C4-C5: 0.10 μF, 400VDC
R1: 1000Ω, 2W, 10%
R2-R3: 1 meg, ½W, 10%
R4: 100 ohm, ½W, 10%
P1: 5 meg, 1 watt potentiometer
T1: U.T.C. PF7 Flash tube trigger transformer
L1: G E FT-106 Xenon flash tube
L2: NE-2 or NE-83 Neon lamp
SCR1: GE C106 D1 or equivalent
SW1: SPST Switch
Misc: Line cord, minibox or enclosure, flashlamp mounting fixture, perforated board.

Fig. P11-1. Schematic and parts list for the xenon strobe light.

repetition rate is set by the rate at which C5 is charged. This charging rate is controlled by potentiometer P1. To change the repetition rate to slower or faster pulsing than the one to five pulses per second, capacitor C5 should be decreased or increased respectively. Rates in excess of eight to ten pulses per second will require additional circuit modifications.

Construction of the strobe can be accomplished in a quite small enclosure since none of the parts are very large, but a larger enclosure provides more room for wiring and mounting the flashlamp. The flashlamp can be mounted as described in Project 3. Remember that the trigger load has pulses of several thousand volts so that high-voltage wire should be used between the trigger lead of the lamp and the trigger-transformer output. Without the isolation transformer the lamp terminals are common with the 120V power line. It is, therefore, doubly important that the lamp terminals be enclosed so that hands and fingers cannot get to them when the unit is in operation.

WARNING

There is medical evidence that stroboscopic lights can have adverse physiological effects on certain people. Caution should be observed in the use and placement of such lighting systems. These systems should never be used in the presence of anyone who has a history of, or potential for, grand mal seizures.

12
Electronic Dice

Many gambling, board, and parlor games use dice. The use of the dice introduces an element of chance (randomness) which increases the interest and excitement of the game. The day of the old ivory cube is coming to an end. Here we present its modern replacement, the electronic die.

The electronic die is built around three ICs, an oscillator, a decade counter, and a BCD decoder. LEDs, a few resistors, a few inexpensive diodes, a capacitor, battery packs, and a switch complete the entire parts list, except for packaging.

The operation of the circuit is straightforward. When the oscillator is running, it applies pulses to the input of the 7490 decade counter, which has been wired to count from zero to five and then reset to zero and count again. The frequency of the counting is high, so that in the counting mode the LEDs appear only to a count zero to five by the 7445 decoder. The signal diodes then provide additional decoding to give the correct display for the die face.

The oscillator is connected independently of the switch so that it will be on for only a brief period while the system is being energized. The switch used is a double-pole double-throw open-frame pushbutton switch. One of the normally-closed contacts of the switch must be bent slightly to insure that the

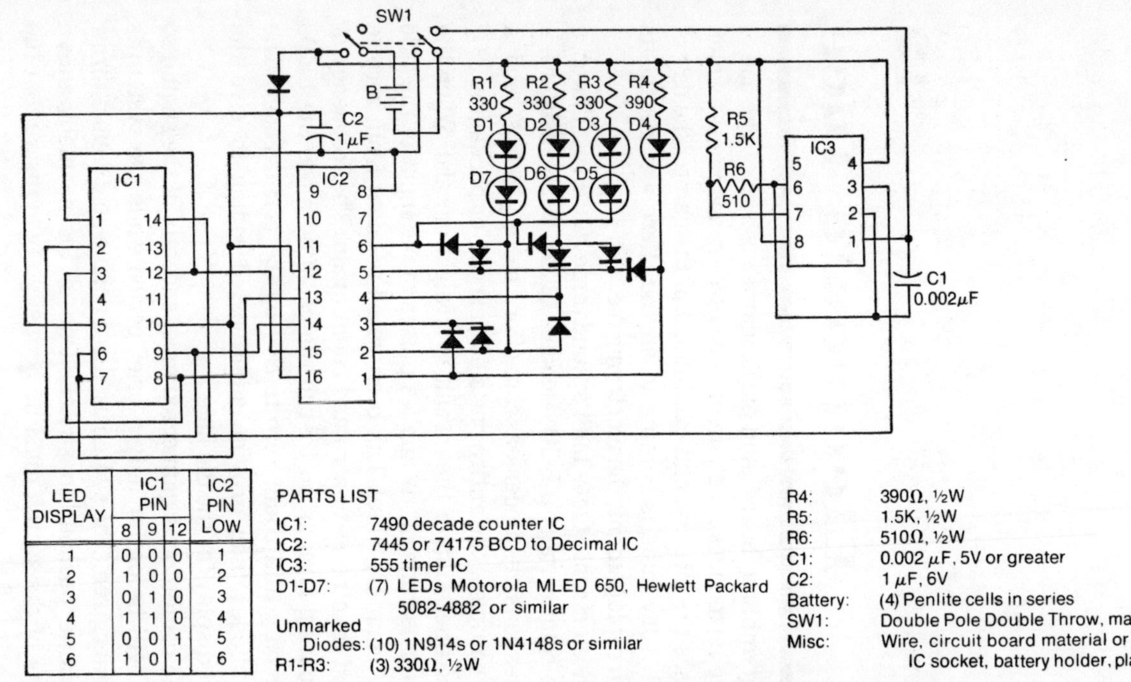

LED DISPLAY	IC1 PIN 8	IC1 PIN 9	IC1 PIN 12	IC2 PIN LOW
1	0	0	0	1
2	1	0	0	2
3	0	1	0	3
4	1	1	0	4
5	0	0	1	5
6	1	0	1	6

PARTS LIST

IC1: 7490 decade counter IC
IC2: 7445 or 74175 BCD to Decimal IC
IC3: 555 timer IC
D1-D7: (7) LEDs Motorola MLED 650, Hewlett Packard 5082-4882 or similar
Unmarked Diodes: (10) 1N914s or 1N4148s or similar
R1-R3: (3) 330Ω, ½W
R4: 390Ω, ½W
R5: 1.5K, ½W
R6: 510Ω, ½W
C1: 0.002 μF, 5V or greater
C2: 1 μF, 6V
Battery: (4) Penlite cells in series
SW1: Double Pole Double Throw, make before break (see text)
Misc: Wire, circuit board material or perforated board, IC socket, battery holder, plastic box

Fig. P12-1. Schematic of the digital electronic die.

normally-open contact closes before the normally-closed opens as shown in Fig. P12-2. If the normally-closed contact is bent too much it will not open when the button is pushed and the oscillator will not stop. The randomness of the die is determined, in part, by how closely you can press the button in the same way each time. But since the oscillator runs at several hundred kilohertz, it is virtually impossible to press the button in an identical way each time, so the randomness of the die is insured.

D1 O   O D2

D4

D3 O   O   O D5

D6 O   O D7

LOGIC TABLE

NUMBER	D4	D1, D7	D2, D6	D3, D5
1	X	0	0	0
2	0	X	0	0
3	X	X	0	0
4	0	X	X	0
5	X	X	X	0
6	0	X	X	X

X=ON
0=OFF

Fig. P12-2. Position the seven LEDs as shown above to give the die the correct display.

The readout for the die is a matrix of seven LEDs. Since six of the LEDs operate only as pairs (for example D1 and D7 are on as a pair for 2 through 6), they have been wired in series to reduce the battery drain. The LEDs when wired as shown in the schematic diagram would be arranged as shown in Fig. P12-3.

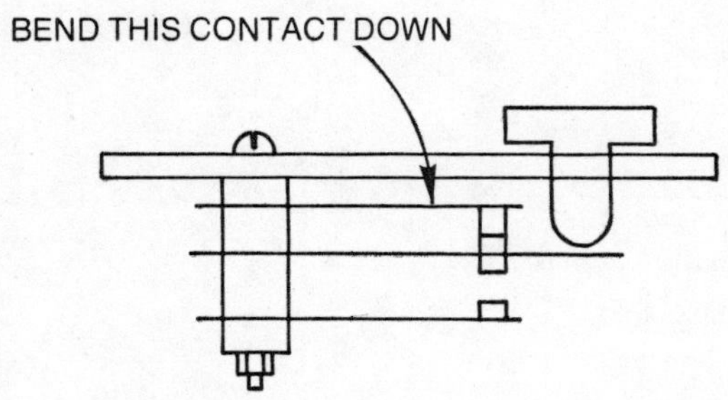

Fig. P12-3. The normally closed contact of one pole of the pushbutton switch must be bent so that the normally open contact closes before the normally closed contact opens.

The circuit can be constructed in any suitably sized box or a 3½-inch plastic cube, such as those used for photographs. With the Plexiglas cube, spray the inside surfaces with white paint and then cut holes in the top for the LEDs plus a hole on one side for the pushbutton switch. The LEDs should be epoxied into position, wired together along with the ballast resistors. Five long leads are then brought out, representing the positive lead and the four leads to the diode logic. The interior foam of the cube is sliced into three equal slices. A one-inch slice of the foam center is inserted into the box. The circuit board with all the components except the switch and battery pack is placed in the center of the second slice of the foam, which has been cut out to allow for the circuitry. Another notch is cut in the second slice, corresponding to the switch location. A pocket in the third foam slice will hold the battery pack. After final connection of the wires the foam is reinserted into the cube, yielding what appears to be a large die.

To check the die for randomness, push the button and record the number. Repeat this operation at least fifty times; the larger the sample the better. Record number of times an odd number appears and the number of occurrences of numbers one to three and four to six.

13 Emergency Lighting System

In many parts of the country, power outages are very common, in other areas they are becoming more common as the power network struggles to keep up with the ever-increasing electrical demands. The emergency lighting system is very handy for those of us who have dark stairways or basements which we must enter during these outages. The system is fully automatic, turning on in about 100 milliseconds after AC power goes off and turning off as soon as power is restored.

The key to the circuit is the biasing of the SCR's gate (see Fig. P13-1). When AC power is supplied to the system, capacitor C1 is charged one-half cycle through D2 and R3. Then on the other half cycle it is discharged through R2. Since R2 is higher in resistance than R3, the net charge on the capacitor is such that the SCR's gate remains with negative bias as long as the AC power is available. When AC power is removed, the capacitor is discharged by the battery and then begins to charge in the opposite direction. When the capacitor's voltage reaches a half volt in the reverse direction the SCR triggers, energizing the lamp.

When AC power is reapplied to the system, the SCR becomes reverse biased by the current through D1 and R1, and

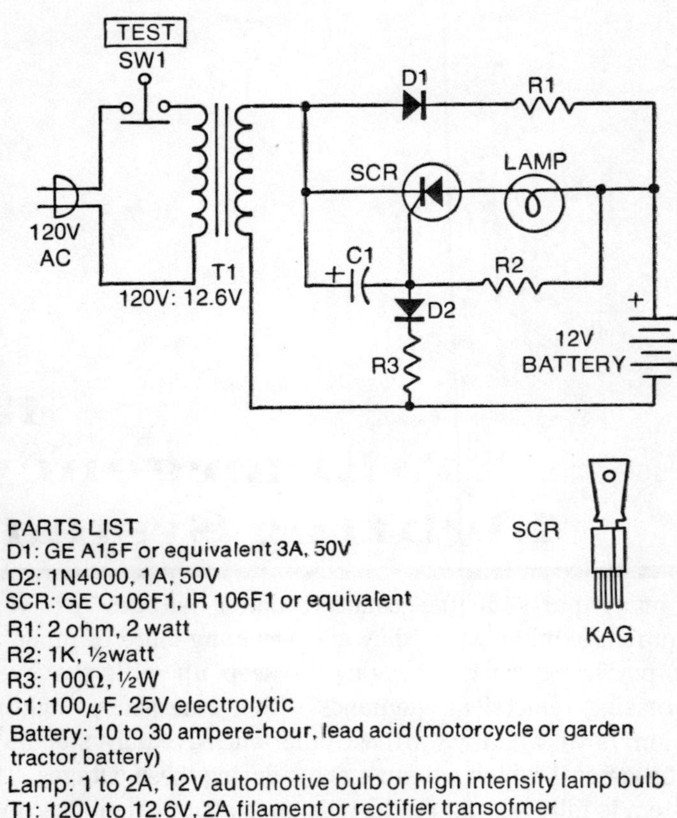

PARTS LIST
D1: GE A15F or equivalent 3A, 50V
D2: 1N4000, 1A, 50V
SCR: GE C106F1, IR 106F1 or equivalent
R1: 2 ohm, 2 watt
R2: 1K, ½watt
R3: 100Ω, ½W
C1: 100μF, 25V electrolytic
Battery: 10 to 30 ampere-hour, lead acid (motorcycle or garden tractor battery)
Lamp: 1 to 2A, 12V automotive bulb or high intensity lamp bulb
T1: 120V to 12.6V, 2A filament or rectifier transofmer
SW1: SPST pushbutton, normally closed
Misc: Oversized battery case, line cord, lamp socket to fit

Fig. P13-1. Basic emergency lighting system schematic and parts list.

at the same time the capacitor charges to reverse-bias the gate. This turns off the SCR and thereby the lamp.

Diode D1 and resistor R1 serve to charge the battery as well. The charging current keeps the battery charged to full capacity so that the system is always ready. The battery size is not critical nor is that of the lamp. The SCR shown in the schematic will handle up to a two-ampere lamp. The bulbs used in the high intensity table lamps are about 1.8 amperes so they will work fine. The battery size depends upon what is available. Garden tractor and motorcycle batteries range

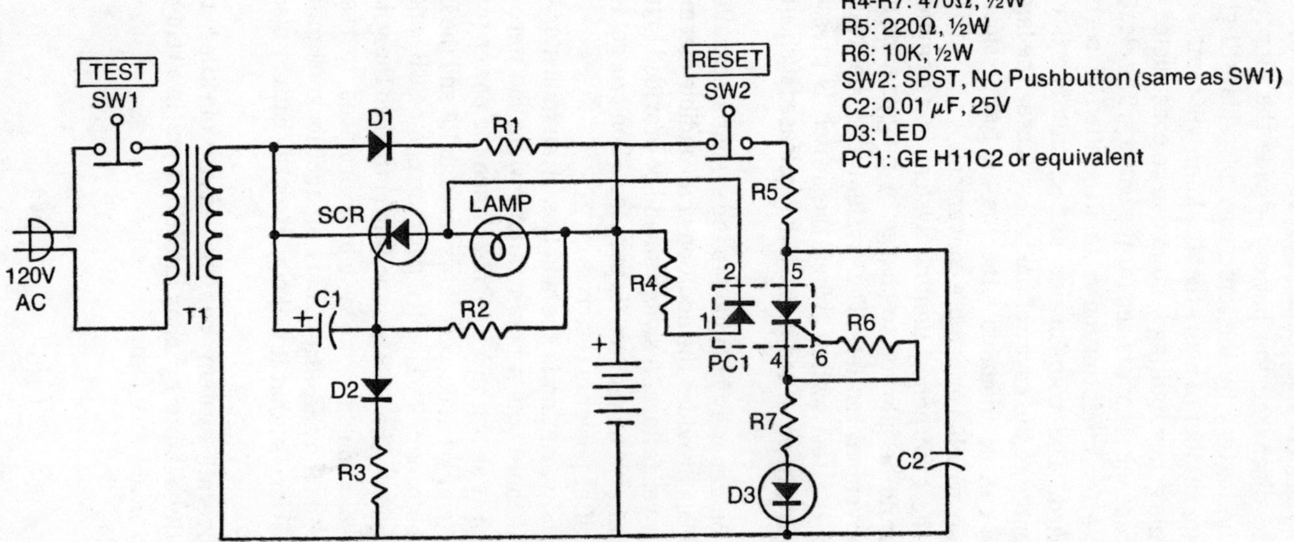

Fig. P13-2. Emergency lighting system with latching indicator. The latching indicator can be used to drive a relay coil which could be connected to sensitive equipment.

from 15 to 30 ampere-hour capacities. To determine how long it will take to discharge the battery, divide the ampere-hour rating of the battery by the lamp current. This will give the approximate number of hours that the lamp will remain on.

The emergency lighting system can be constructed within a battery case that is oversized for the battery. The electronics are suitable for mounting on a small section of perforated circuit board. The transformer and circuit board can be screw-mounted to the cover of the battery case. The lamp and test switch can be mounted either on the top or side of the battery case, whichever is more convenient.

After the emergency lighting system is assembled, the battery should be allowed to charge by plugging the system into a convenient outlet. The system can be tested by depressing the test switch. When the switch is pressed the lamp should turn on, when the switch is released it should turn off.

A 12V buzzer may be added to the system in parallel with the lamp. This provides the addition of an audible alarm in the case of a power failure. A switch should be provided to turn off the buzzer after it has served its purpose. (Don't forget to reset the switch after power is restored.)

A second variation is the addition of an indicator, which will show if there was a power failure while you were away. This system is shown in Fig. P13-2. When the power fails and the SCR is triggered, current flows through R4 and the LED of the SCR photocoupler. This triggers the photo-SCR, and LED D3 is energized. The SCR will not turn off when the voltage is restored, so a pushbutton switch is provided to reset the indicator. This circuit draws only an additional 50 milliamps from the battery so that it will not affect the battery discharge time.

The overall system can be easily modified to run higher wattage lamps by using the SCR to pull in a relay which would turn on any size lamp desired.

14 Electronic Flame-Effect

There are many homes with artificial fireplaces. Scout troops like to have their campfires, but cannot in most meeting places. The wind keeps blowing out the candle in the jack-o-lanterns. The previous owners electrified that old kerosene lantern and it just doesn't seem the same. All of these situations are ideal candidates for an electronic flame-effect circuit.

The circuit shown in Fig. P14-1 creates the flame effect using a three-way bulb and a simple circuit. The key to the circuit is the two neon lamp relaxation oscillators. These two oscillators are connected such that each modifies the frequency of the other, creating a rather random effect.

The three-way lamp is normally in a light fixture connected so that the common point is connected to one side of the AC line. The other two terminals are then switched so that either one or both filaments are energized. By connecting the filaments in series, as in this circuit, each filament sees a reduced voltage. This gives the lamp a reddish color rather than the white color we normally expect. When the SCR triggers, it puts a short across the 30-watt filament and thereby increases the voltage across the 70W filament. This changing of voltage causes the 30W filament to dim and the 70W filament to brighten. If instead of triggering the SCR at

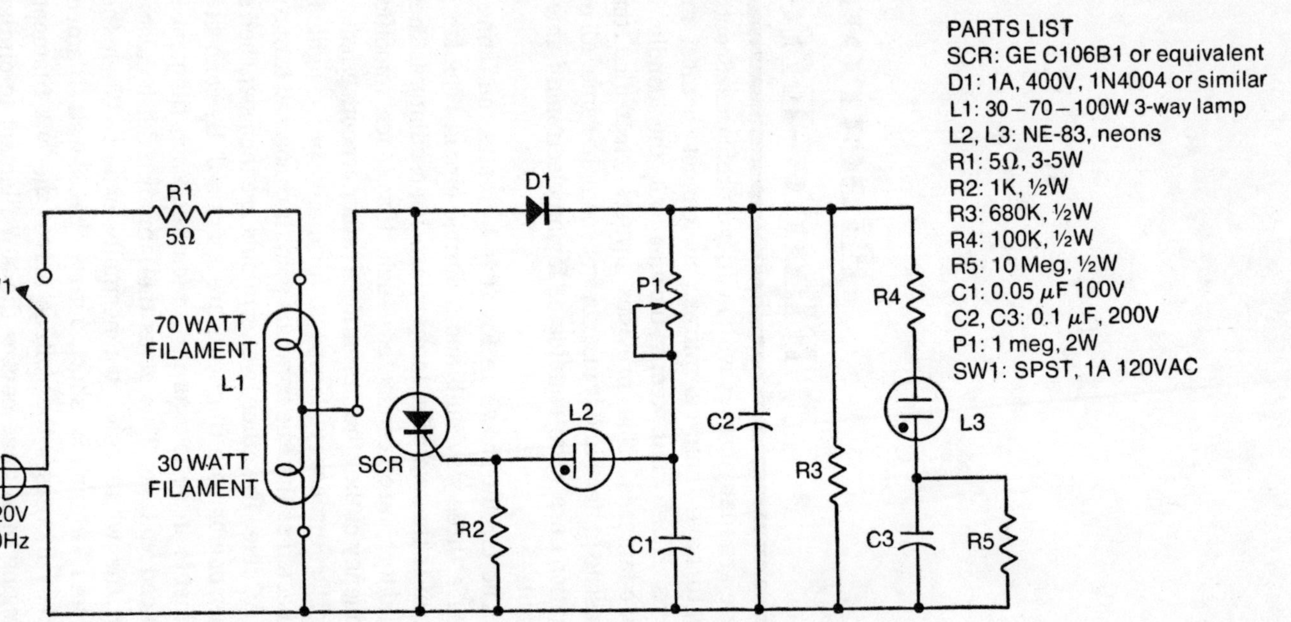

Fig. P14-1. Electronic flame-effect circuit schematic and parts list.

the same phase angle of the AC line each cycle, the phase angle was to be varied across the full (or at least most of the) sine wave, then the light would flicker rather than just dim. This is in effect what the two neon oscillators cause to happen.

The one oscillator consists of C2, R4, L3, and C3 in parallel with R5. When the line voltage is positive, C2 charges to the peak of the AC voltage if the SCR is off. If the voltage on C3 is such that the difference in the two capacitor voltages (C2 and C3) exceeds the neon breakdown voltage, neon L3 turns on and discharges C2 into C3. If the AC line voltage is high then both capacitors become charged. When the neon is off, C3 discharges slowly through R5. When the line voltage is low, or the wrong polarity, or when the SCR is on, C2 discharges through R3, P1, and possibly R4 as described above.

Now, if you're not completely confused, the second oscillator, consisting of P1, C1, L2, and R2 runs off of the first. Capacitor C1 charges through P1 from the stored energy of C2. The rate of charge of C1 is varied because of the varied voltage on C2. When C1 reaches the breakdown voltage of L2, the capacitor partially discharges through the lamp and triggers the SCR.

The construction is left to the reader since the final use will necessitate the form that the unit will take. The author's was used for the jack-o-lantern on a one-shot deal, so I did not bother with switches, sockets, or even a circuit board but just wired the circuit on the lamp itself using some rather heavy bus wire. Extra care should be observed if this method is used. Once the circuit is built, fiddle with the pot. By varying its resistance different effects can be obtained. The addition of some slightly crinkled aluminum foil, when using the flame-effect circuit as a fire, will add additional effects.

15 Low-Cost Stage-Lighting Control

With the many amateur theater groups, rock bands, and small clubs around today, most cannot afford the expense of the stage-lighting panels used in major theaters. This lighting system is inexpensive, completely portable, and capable of controlling any 120V spotlight up to 1800 watts.

The system as shown in Fig. P15-1 is set up to control one lamp but is easily expandable to several lamps. The system contains presets for the level and face rate for as many scenes as desired. The control center operates from low voltage with only an isolated low-voltage control wire running to the light. This can be located at any distance from the control panel.

The power control unit is mounted in a separate box which is placed adjacent to the lamp. (See Fig. 15-2.) It contains a triac phase-control circuit, similar to those found in lamp dimmers. The main modification is the replacement of the potentiometer with a photocell. The photocell's resistance is varied by a low-voltage lamp placed adjacent to the photocell. The lamp's light output is controlled by the control center.

The control center is a transistor emitter-follower circuit. The lamp in the power control unit is the load for the emitter follower. The level of the light output is controlled by resistor dividers R3, R4, and R5; the rate of change from one scene to

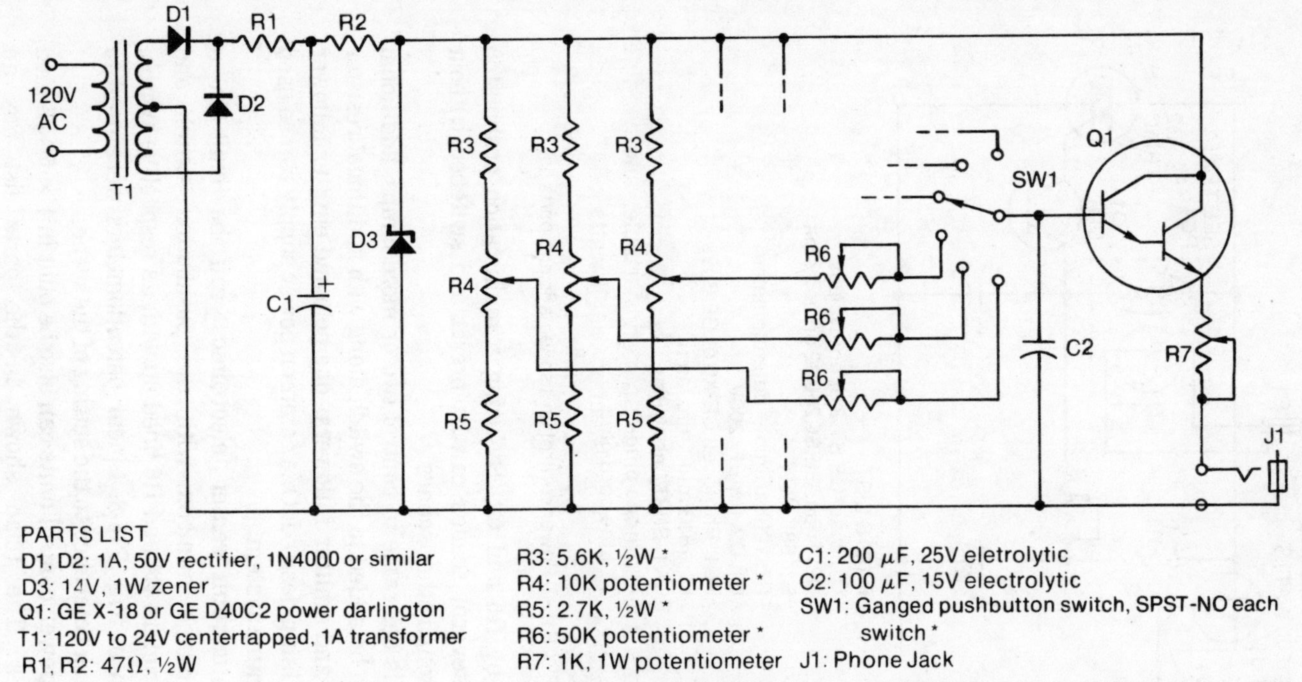

PARTS LIST
D1, D2: 1A, 50V rectifier, 1N4000 or similar
D3: 14V, 1W zener
Q1: GE X-18 or GE D40C2 power darlington
T1: 120V to 24V centertapped, 1A transformer
R1, R2: 47Ω, ½W
R3: 5.6K, ½W *
R4: 10K potentiometer *
R5: 2.7K, ½W *
R6: 50K potentiometer *
R7: 1K, 1W potentiometer
C1: 200 μF, 25V eletrolytic
C2: 100 μF, 15V electrolytic
SW1: Ganged pushbutton switch, SPST-NO each switch *
J1: Phone Jack

*Number of poles on switch equals number each of R3, R4, R5 and R6.

Fig. P15-1. Control center schematic and parts list.

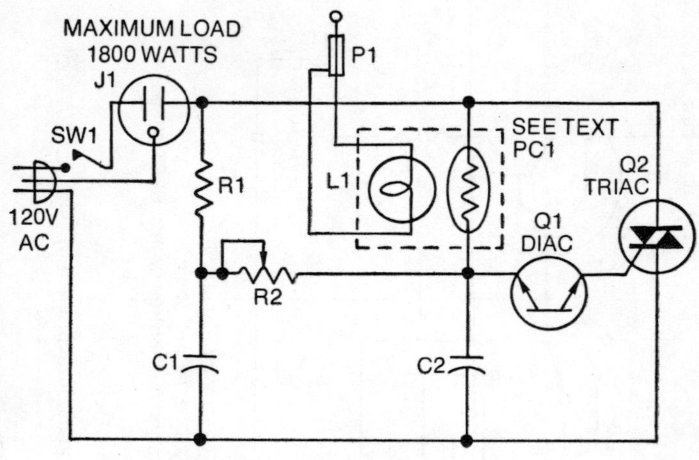

PARTS LIST:
Q1: Diac, GE ST-2 or equivalent
Q2: Triac, GE SC260B2 on heatsink
R1: 68K, ½W
R2: 100K trimmer potentiometer
C1, C2: 0.1 μF, 200V
PC1: Photocell GE-X6 or GE B425
L1: 1813 miniature lamp
SW1: SPST line switch
P1: Phone plug
J1: Line cord plug

Fig. P15-2. Power control unit schematic and parts list.

another by R6 and C2. By using a multistation pushbutton switch, several scenes can be preset and switched without going through other scenes.

If it is desirable to control two or more lamps, additional poles can be added to the switch along with additional resistor dividers and emitter followers; one pole and emitter follower for each lamp desired. One common power supply can be used for the entire system.

The control center electronics can be built in a 4 × 5 × 6-inch minibox with the pushbutton switch and potentiometers R4 and R6 lined up with its respective switch, as shown in Fig. P15-3. Linear potentiometers were used to allow quick reference to the setting of the scene.

The power control units can also be built in 4 × 5 × 6-inch miniboxes. The triac shown in the parts list has an

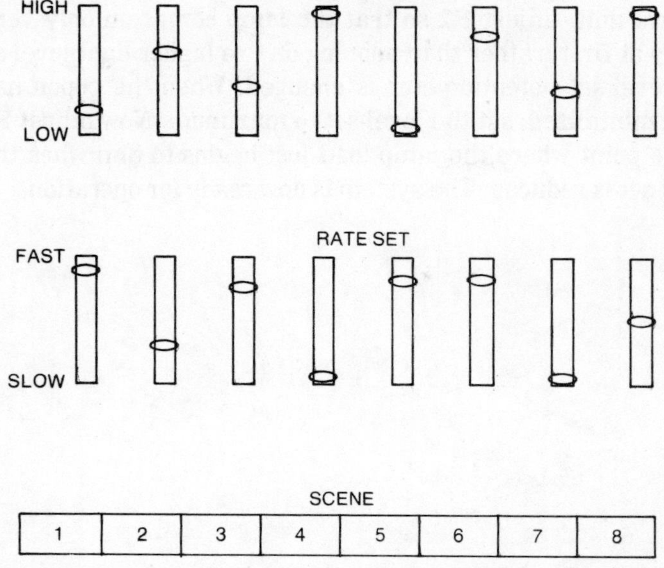

Fig. P15-3. Layout for controls on control center using linear potentiometers. The two potentiometers immediately above the switch station control that station.

electrically isolated stud. The heat sink for the triac can be a 4 × 8 × ⅛-inch aluminum plate bent into a *U* shape. This plate should be bolted to the outside of the minibox. A quarter-inch diameter hole should be drilled through the center of the plate and the box. The triac is mounted from the inside of the box with only the end of the isolated stud and the nut exposed on the outside. The balance of the power control unit can then be assembled inside the box on either terminal strips or a small section of perforated board.

The lamp and photocell are used as a photocoupler. It is therefore necessary to assemble them together. The lamp should be placed directly on the lens of the photocell and the two should then be completely wrapped in black electrical tape so that no external light can enter the photocell.

After the two units are assembled, a one-time adjustment of the system must be made. Connect the two units and use any convenient lamp for a load. With both units plugged into 120 VAC, adjust R7 such that the lamp goes completely off when

R4 is changed from maximum to minimum. In the power control unit, adjust R2 so that the lamp comes on only very dimly at first, rather than popping on to a higher light level as the level-set potentiometer is changed. When the popon has been minimized, set the level-set to maximum. Now adjust R7 to the point where the lamp load just begins to dim when the level-set is reduced. The system is now ready for operation.

16
Multiple-Scale Light Meter

Most photographic hobbyists have at least one light meter in their kit, many have several. Here is an inexpensive light meter for those who don't have one or just would like to have another.

The system is quite simple. A photovoltaic cell provides the energy to drive a DC microammeter. A four-position divider is provided to change the light meter range.

The selenium photovoltaic cell must be compensated to approximate the response of the eye. This can be accomplished in the simplist manner by the use of the Kodak Wratten filter listed in the parts list. If this filter is not used, the meter will indicate 21% higher in noon sunlight than it will for the same light level from an incandescent bulb (operated at 2750°K).

The assembly of the light meter depends upon the meter being modified. Since microammeters can be purchased at costs from a couple of dollars to some that run over fifty, scout around and see what is available in the junk box or from friends first. An old 20K per volt VOM is one source. The resistors are easily assembled directly on the switch, leaving only the photovoltaic cell and filter to be mounted.

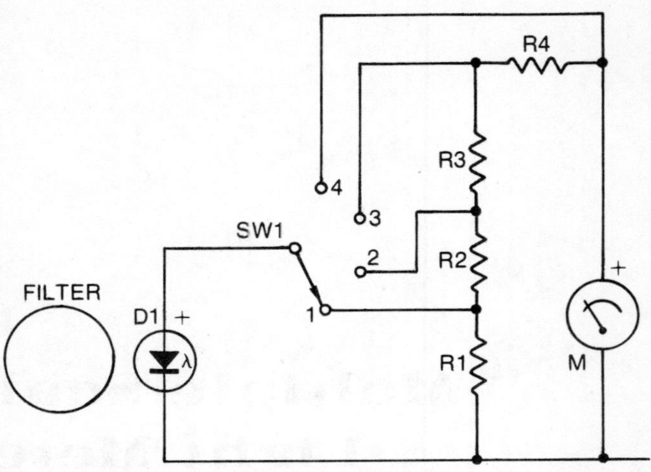

PARTS LIST
D1: International Rectifier B-30 selenium photovoltaic cell
R1, R2: 150Ω, 1%
R3: 300Ω, 1%
R4: 600Ω, 1%
M: 0 – 50 μA meter (internal resistance of less than 2500 Ω)
Filter: Kodak Wratten 102 Filter
SW1: Four position, single pole rotary switch

Fig. P16-1. Multiple-scale light meter parts list and schematic.

The assembled unit is virtually useless until it is calibrated. To calibrate the light meter, a precalibrated light meter is required. Remember a calibrated light meter will only be as good as the standard to which it is calibrated, so try to borrow the best light meter you can get for the calibration procedure. Using any fixed light source, such as a standard incandescent lamp, measure the light level with the standard light meter. Move the light source closer or farther away to get a nice even reading; four or five footcandles would be best. With the switch in the number four position, place the photovoltaic cell with its filter in the exact position of the standard meter's sensor. The meter will probably read over full scale. With some flat-black paint, carefully paint a frame around the edge of the photovoltaic cell until the meter reads full scale. By switching the meter to position three, the needle should drop to midscale, at two, quarter scale, and at one, one-eighth of full scale. The scales on the meter are then: four is full scale as calibrated, three is twice, two four times, and nine is eight times the range. If you started with five footcandles, the scales are 5, 10, 20, and 40 respectively.

17 Optoelectronic Key Lock

Most of us, at one time or another, have wished that we had a lockset that was unique to ourselves. One that no hairpin or pick could open. The optoelectronic key lock is just that.

This project uses the solid-state relay, as described in Chapter 5, to drive a solenoid. The solid-state relay is controlled by a photocoupler transistor circuit. The photocouplers used are of the interrupter type. These units have a 0.12 by 0.30-inch slot between the LED emitter and the transistor detector. When the slot is empty, or filled with an infrared transmissive material, the phototransistor will conduct current in response to the light of the LED. If the slot is filled by an opaque material, the transistor will remain off. The entire slot isn't sensitive. The sensitive part is only a small sixty-mil area within the gap. If an object is placed in the slot with a sixty-mil hole in the correct location, the LED light will get through to the transistor.

The key lock uses an array of these interrupter photocouplers which are mixed so that only the correct "light combination" will open the lock. With 4 photocouplers there are 16 combinations, 5 gives 32, and so on.

The electronics are shown in Fig. P17-1. As mentioned, the solid-state relay drives the solenoid; the relay is turned on by

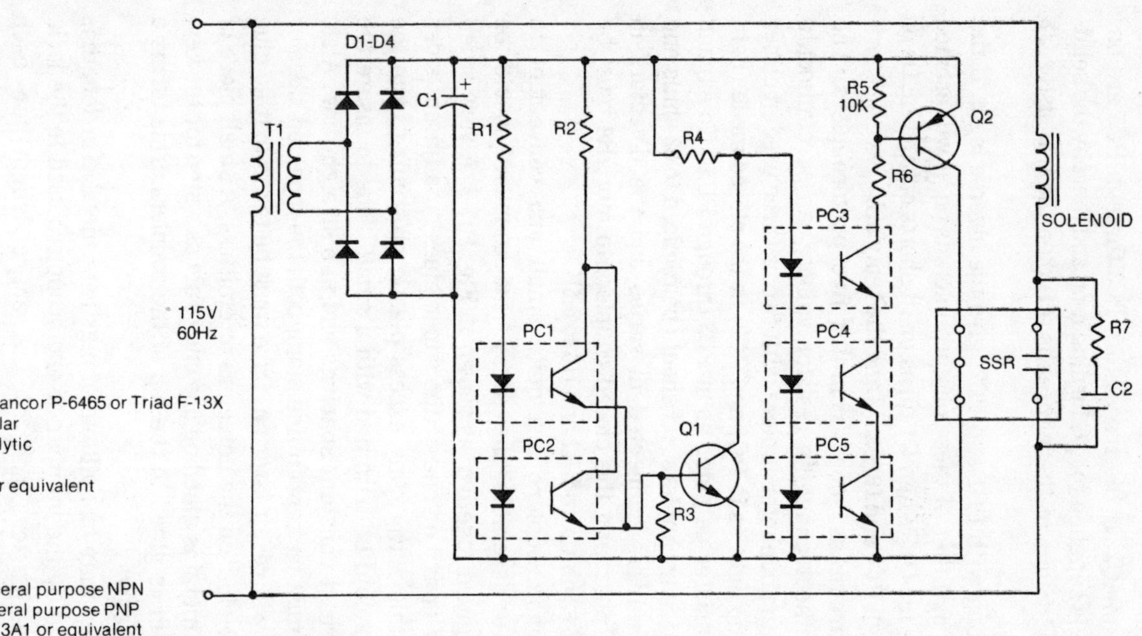

PARTS LIST
T1: 115 to 6.3V, .6A: Stancor P-6465 or Triad F-13X
D1-D4: 1N4000 or similar
C1: 500μF/10V electrolytic
C2: .05 μF/200V
SSR: Crydom D1202 or equivalent
R1, R4: 120Ω, ½W
R3, R5: 10K, ½W
R2, R6: 6.8K, ½W
R7: 470Ω, ½W
Q1: 2N5172 or any general purpose NPN
Q2: 2N5354 or any general purpose PNP
Photocouplers: GE H13A1 or equivalent
Solenoid: Any 120V unit up to 1A coil current such as Guardian 2HD-120VAC

Fig. P17-1. Schematic and parts list for the optoelectronic key lock. The number of photocouplers can be from two to six. The text explains how to vary the number. The schematic shows five in use.

the transistor Q2. If the transistors PC3 through PC5 are on, base drive is supplied to Q2, turning it on and thereby the solid-state relay. If one of these three is off, then the transistor Q2 will be off.

Q1, along with PC1 and PC2, give the inverse logic. If the transistor of either of these photocouplers is on, it will provide base drive for Q1. When Q1 turns on, it shorts out the LEDs of PC3 through PC5 and turns Q2 and the relay off.

The circuit as shown contains five photocouplers. The circuit, without changing resistor values, can accommodate from four to six photocouplers with two or three used in either the open (no light as in PC1 and PC2) or the closed mode (light on the transistor detector as in PC3 through PC5). To expand the open to four couplers, R1 should be reduced to 82 ohms and all four emitters connected in series and the detectors in parallel. To expand the closed couplers to four, R4 must be reduced to 82 ohms and R2 and R6 to 5.6K.

The photocouplers are quite small and convenient to handle. To assemble the array, the couplers should be carefully glued together as shown in Fig. P17-2. A very small amount of epoxy or one of the super glues on adjacent sides will hold this array. Any excess plastic protruding from the couplers should be trimmed with a small blade to insure the centers are at constant spacings and in a straight line. Also, the slots must be kept in line along both the tops and sides or problems will result later on. Care must be taken that no glue gets in the slots on the detectors or emitters. A small piece of plastic or metal is glued on one end to act as a stop for the key, and a cover is glued to the top of the couplers. This forms a tunnel or keyway.

The key may be built out of plastic or metal. The length is determined by the number of photocouplers used in the lock. Allow one-quarter inch of the entire length for each photocoupler plus two inches for a handle. If the key is cut precisely, it will slide into the tunnel with very little up and down play. At this point the combination is determined. Drill a 1/16-inch diameter hole (only at the centers shown in Fig. P17-2 corresponding to the photocouplers used in the closed mode). The hole placement must be accurate. If you do not

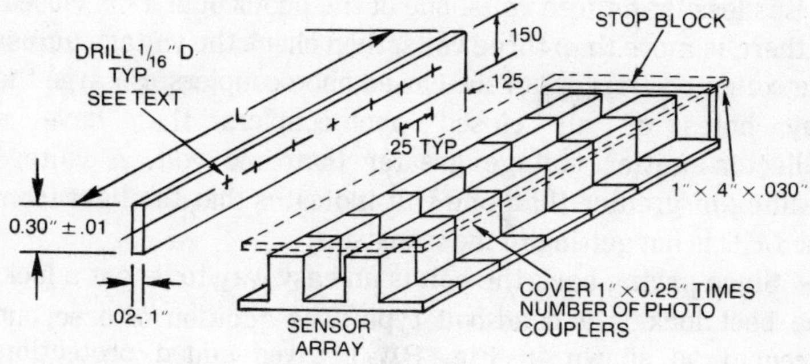

Fig. P17-2. The photocoupler array is assembled as shown here. The dimensions of the key are shown also.

have the proper tools to accurately locate the holes, use a larger drill bit to insure that the light path is clear.

The electronics can be assembled within a 4×5×6-inch box. The photocoupler array can be mounted on a perforated circuit board using several of the holes on the photocouplers. The open end of the photocoupler tunnel should be placed at the edge of the box with a hole large enough for the key to enter the tunnel. When making the photocoupler display, make sure that the connectors of the normally open and normally closed couplers are not interchanged.

Once the unit is assembled, inserting the key should energize the solenoid. If not, check the collector voltage of Q1.

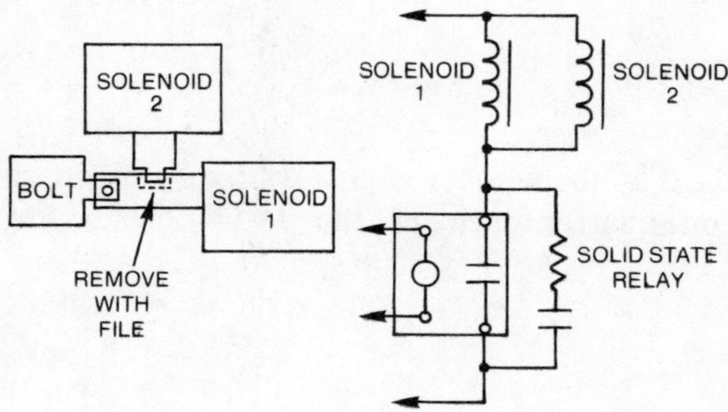

Fig. P17-3. For added security a second solenoid may be added as shown here. The second solenoid gives an action similar to a dead bolt.

If it is less than three volts, one of the photocouplers is closed. If there is more than three volts, then check the voltage across the collector-emitters of the closed photocouplers. Enlarge the key holes of all closed photocouplers that have a collector-emitter voltage greater than one volt. A voltage reading of greater than one volt indicates that the light from the LED is not getting to the transistor.

Since prying back the bolt is an easy way to defeat a lock, the best lock is a dead-bolt type. The addition of a second solenoid as shown in Fig. P17-3 gives added protection. Without the second solenoid the bolt can be pushed back against the solenoid return spring. The second solenoid at right angles locks the bolt. The bolt can not be pushed in unless the second solenoid is energized.

Index

Index

T

V

W